To
Shyamali, Somita, Soumen and Swagota

This book is an undergraduate textbook for students of electrical and electronic engineering. It is written at an intermediate level, with second year students particularly in mind, and discusses analogue circuits used in various fields, including the interfacing of microcomputer systems. Basic electronics has been omitted so that appropriate emphasis can be given to the design of the most popular and useful circuits. Indeed, the contents of Chapters 3, 5, 7 and 8 are not covered together by any other single textbook available on the market.

The author begins with a summary of the knowledge which is prerequisite to a full study of the book. The second chapter then discusses the operation of several basic circuits which are commonly used in most integrated circuit chips. Other sections included in this chapter will acquaint the reader with different types of power amplifier circuit. The three important areas on which the book places greatest emphasis are operational amplifier circuits and their applications, data acquisition circuitry, and computer aided analysis and design. Oscillators, phase-locked loops and different types of modulation are also discussed, and particularly helpful detail is given to the important topics of phase-locked loops and analogue filter circuits. Each chapter contains a significant number of worked examples, and several carefully chosen problems at various levels of difficulty are also included to help the reader gain a better understanding of the topics under discussion.

Each topic has thus been carefully selected, and the author concentrates on the practical details and applications of the material he covers. Both students and practising engineers alike will therefore find this book extremely useful and informative.

Electronics texts for engineers and scientists

Editors

P.L. Jones, *University of Manchester*

P. J. Spreadbury, *University of Cambridge*

Analogue electronic circuits and systems

Analogue electronic circuits and systems

AMITAVA BASAK

University of Wales College of Cardiff

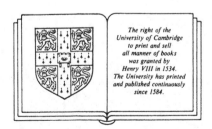

The right of the
University of Cambridge
to print and sell
all manner of books
was granted by
Henry VIII in 1534.
The University has printed
and published continuously
since 1584.

CAMBRIDGE UNIVERSITY PRESS

Cambridge

New York   Port Chester   Melbourne   Sydney

Published by the Press Syndicate of the University of Cambridge,
The Pitt Building, Trumpington Street, Cambridge CB2 1RP
40 West 20th Street, New York, NY 10011-4211, USA
10 Stamford Road, Oakleigh, Melbourne 3166, Australia

© Cambridge University Press 1991

First published 1991

British Library cataloguing in publication data

Basak, Amitava
  Analogue electronic circuits and systems.
  1. Circuits integrated analogue, analysis and design
  I. Title   II. Series
  621.3815

Library of Congress cataloguing in publication data

Basak, Amitava
  Analogue electronic circuits and systems / Amitava Basak.
      p.   cm. — (Electronics texts for engineers and scientists)
      Includes bibliographical references.
      ISBN 0 521 36046 3 (hardback). — ISBN 0 521 36913 4 (paperback)
      1. Analog electronic systems.   I. Title.   II. Series.
   TK7870.B357   1991
   621.381'5–dc20
                                                              90–21775
                                                                   CIP

ISBN 0 521 36046 3 hardback
ISBN 0 521 36913 4 paperback

Transferred to digital printing 2003

VN

Contents

Contents

Preface

This book is intended particularly as a text for undergraduate students of electrical and electronic engineering at the intermediate level of a degree course. Some material may, however, be appropriate to the final year. Each topic has been deliberately selected and emphasis has been given to operational amplifier circuits and their applications, data acquisition circuitry and computer aided analysis and design. Other useful topics which have been covered in some detail are analogue filter circuits and phase-locked loops. A list of the prerequisite knowledge for this text is given in the first chapter.

This book is most appropriate for students because (i) a specific subject area, analogue circuits for more advanced students has been highlighted and (ii) particular attention has been given to a descriptive treatment of practical details and applications, e.g. CAD rather than theoretical analysis, because these areas are often neglected and are essential for practising engineers. Therefore it is hoped that students and engineers alike will find this text useful and informative, but less analytical than many other books presently available, and thus be able to cover a wider range of topics within a given period of time. Several carefully chosen problems set at various levels of difficulty are included at the end of each section to help readers gain a better understanding of the topics under discussion.

I would like to thank my parents and my immediate colleagues both academic and industrial for their encouragement and advice in the preparation of the manuscript. Thanks are also due to the staff of Cambridge University Press for their support through the many stages of production. I am also grateful to my research assistant Paul Ling and student Fatih Anayi for checking the problems.

1

Introduction

In recent years there has been rapid progress in electronic circuit design and the main reason for this is the advance in digital techniques. This volume differs from the texts which are available on the market nowadays in two respects. Firstly it covers only analogue electronic circuits and systems; secondly basic electronics is omitted so that appropriate emphasis can be given to the design of the most popular and useful analogue electronic circuits. The following are prerequisites for studying this text:

(a) P-N junction diodes: principles of operation both in the forward and reverse mode, characteristic equation, resistance and junction capacitance, Zener diodes.

(b) Junction transistors: principle of operation, common-emitter (CE), common-collector (CC) and common-base (CB) configurations, static characteristics, definition of active, cut-off and saturation regions, the concept of load lines and the need for biasing, the transistor as an amplifier.

(c) Amplifiers: voltage and current gains (A_v and A_i), input and output resistance (R_{in} and R_{out}), frequency response concept, the use of the h-parameter model of the transistor for circuit analysis, midband frequencies of the CE, CC and CB configurations and calculation of A_v, A_i, R_{in} and R_{out} for each case.

(d) Field effect transistor: principle of operation, static characteristics, load lines, biasing circuits, use as an amplifier.

(e) Positive and negative feedback and their advantages and disadvantages.

(f) Operational amplifiers: ideal amplifier, analysis of inverting, noninverting, differential, buffer and summer amplifiers, use of operational amplifiers as integrators and differentiators.

A list of books which comprehensively cover the above topics is given at the end of this volume. Standard symbols are used throughout the text and a glossary is included as one of the appendices.

In the second chapter the operation of several basic circuits which are commonly used in most integrated circuit chips is described. A section is also devoted to the analysis of multistage amplifiers and the ways of choosing the right configuration of transistor amplifiers for a particular stage. These are included in the text in order to acquaint the reader with the operational amplifier integrated circuit which is the building block of most analogue circuits. The principle of operation of tuned amplifiers is briefly explained in the penultimate section. In the last section, different types of power amplifiers are studied. Various types of heat sinks and their use in power circuits are also discussed in this section.

Chapter 3 deals with operational amplifiers in great detail, but at a level higher than the introductory one which is [one of the] required knowledge, as mentioned earlier. The circuits are analysed and designed assuming operational amplifiers to be ideal, but in practice they are not so. In this book, therefore, the chapter starts with the imperfections in operational amplifiers, their effects on various operational amplifier circuits and the ways in which readers can minimise these effects. Several widely used linear and nonlinear circuits using operational amplifiers are discussed in the remaining sections of this chapter. One of the major applications of operational amplifiers is in active filters and therefore readers will find both resistor–capacitor and switched-capacitor type active filters, which are discussed in this chapter with design examples, rather useful. Principles of design of waveform and function generators and also analogue computation using operational amplifiers are comprehensively studied in the next section.

The next three chapters describe oscillators, phase-locked loops and different types of modulation respectively. Principles of design and operation of oscillators using transistors, operational amplifiers and quartz crystals have been presented. The most common uses of phase-locked loops in the field of communications and also in the field of control of motor speed are discussed to a limited extent, with a full treatment of the theory of phase-locked loops which will help readers to understand and design similar circuits.

In Chapter 7, under the heading of data acquisition and distribution systems, analogue-to-digital and digital-to-analogue conversion techniques are discussed in great detail. Sample-and-hold circuits, multiplexers and demultiplexers are also studied in great depth. Analysis of errors in individual circuits and also in complete systems are discussed. In the age of computerised measurement in research and in most aspects of control systems this chapter is a significant part of the text.

The last chapter deals with computer aided analysis and design of electronic circuits. It describes various computer aided design models giving special attention to the Ebers–Moll model of a transistor. Models for bipolar junction and field effect transistors and also integrated circuits are described. Techniques of a.c. small signal, d.c. and transient analysis of circuits are discussed with the help of examples. Several commercially available computer programs are also discussed. Two widely used software packages are described with the aid of circuits and their analyses in order to make students familiar with the procedure for drawing and analysing various active and passive circuits.

To summarize it can be said that this volume on analogue electronic circuits and systems has been written mainly keeping in mind the requirements of undergraduate students at the intermediate level. Practising engineers interested in various aspects of analogue electronic circuit design will also find this text informative.

2

Transistor circuit techniques and amplifiers

Objectives

At the end of the study of this chapter a student should be:

1. familiar with the operation of the differential amplifier, its voltage gain, common-mode gain and common-mode rejection ratio
2. familiar with constant current sources and current mirror circuits
3. capable of explaining the principle of Darlington connections
4. able to design level shifting circuits
5. familiar with multistage amplifiers and able to calculate their input and output impedances and overall current and voltage gains
6. able to design class A, class B and tuned amplifiers
7. familiar with different types of heat sinks and able to choose the right heat sink for a particular circuit

2.1 Linear integrated circuits

Complete multistage amplifiers and other linear devices can be constructed on a single chip of silicon occupying a very small volume by using modern techniques for the fabrication of integrated circuits. In the case of monolithic integrated circuits, all components may be manufactured on the chip by a diffusion process. A diffusion isolating technique is used to separate the various components from each other electrically. The design techniques used for the construction of these integrated circuits are basically the same as those used to build circuits employing discrete components, although, in many cases some modification in techniques is needed.

The operational amplifier is the most common type of integrated circuit (small scale integration) which is widely used with different forms of external circuitry to build summers, subtractors, integrators, filters, etc. Audio amplifiers, timers, modulators and frequency dividers are only a few among many other types of integrated circuits.

In this section, the operation of several basic circuits which are commonly used in integrated circuits are described.

Fig. 2.1. Basic differential amplifier.

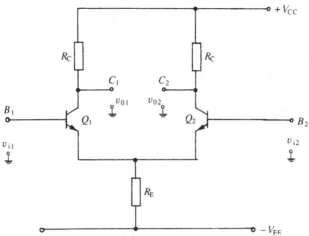

2.1.1 Differential amplifiers

One of the basic circuits used in operational amplifiers is the differential amplifier. This forms the basis of practically all operational amplifiers. Also known as the long tailed pair, one of its advantages is that its gain tends to be very stable if any variations in supply voltage or ambient temperature occurs.

The basic form of the differential amplifier is shown in Fig. 2.1. It consists of two identical bipolar transistors coupled at their emitters. The collector resistors, connected between the positive voltage supply rail and the collectors of the transistors, are of the same value, R_c. Two input signals are applied to the bases of the two transistors Q_1 and Q_2 with respect to ground and the output is usually obtained from either of the two collectors, again with respect to ground.

Differential voltage gain

The differential voltage gain, A_d is the ratio of the output voltage to the difference between the input voltages applied to terminals B_1 and B_2. If the output voltage is obtained from either of the two collectors with respect to ground, then the ratio gives the single-ended differential voltage gain $A_{d(s)}$. If the output voltage is the difference between the two collector voltages then the ratio yields the double-ended differential voltage gain $A_{d(d)}$.

In order to find the single-ended differential voltage gain $A_{d(s)}$ of the circuit let us apply a very small voltage v_d between the two bases B_1 and B_2. Since the circuit is fundamentally symmetrical and the total emitter current

Fig. 2.2. Small signal equivalent of the differential amplifier circuit.

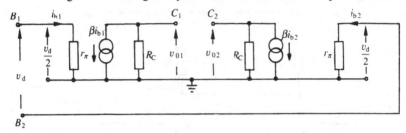

is constrained to be constant by a large R_E, we may assume that the potential of the base B_1 rises by $v_d/2$ while that of the base B_2 falls by $v_d/2$ thus increasing the collector current of Q_1 by a small amount and at the same time reducing that of Q_2 by an equal amount. The voltage across R_E remains constant. Figure 2.2 shows the small signal equivalent of the differential amplifier circuit in which the resistor R_E has been omitted since the voltage across it is constant. From the figure we may write

$$i_{b_1} = -i_{b_2} = \frac{v_d}{2} \bigg/ r_\pi = \frac{v_d}{2r_\pi} \tag{2.1}$$

therefore the output voltage at terminal C_1 is given by

$$v_{o1} = -\beta i_{b1} R_c = -\beta R_c \frac{v_d}{2r_\pi} \tag{2.2}$$

where β is the common emitter current gain, and r_π is the base-emitter resistance.

From (2.2) the single ended differential voltage gain is therefore,

$$A'_{d(s)} = \frac{v_{o1}}{v_d} = -\frac{\beta R_c}{2r_\pi} = -\frac{g_m R_c}{2} \tag{2.3}$$

where g_m is the transconductance.

Similarly we may find that the single-ended differential voltage gain at terminal C_2 is

$$A''_{d(s)} = \frac{v_{o2}}{v_d} = \frac{-\beta i_{b2} R_c}{v_d} = \frac{\beta R_c v_d}{v_d 2r_\pi} = \frac{g_m R_c}{2} \tag{2.4}$$

We observe from (2.3) and (2.4) that the voltage gains obtained at the two collectors are identical in magnitude but differ in phase by 180°.

The double-ended differential voltage gain is given by

$$A_{d(d)} = -\frac{g_m R_c}{2} - \left(\frac{g_m R_c}{2}\right) = -g_m R_c \tag{2.5}$$

Thus the double-ended differential voltage gain is twice the single-ended differential gain.

Common-mode gain

The common-mode gain, A_c, is defined as the ratio of the output voltage at either of the two collectors to the input voltage applied simultaneously to terminals B_1 and B_2. If we now apply the same voltage v_c to both the bases of transistors Q_1 and Q_2 then the transistors with their respective loads are effectively in parallel. The small signal equivalent for the circuit will be as shown in Fig. 2.3. In this case the current flowing through the resistor R_E changes with the variation in v_c and thus remains in the equivalent circuit. The latter can be further simplified to the circuit shown in Fig. 2.4. We may write from this figure

$$v_{o1} = v_{o2} = -\beta R_c i_b$$

and

$$v_c = r_\pi i_b + 2(\beta + 1)R_E i_b$$

Fig. 2.3. Equivalent circuit for common-mode voltage gain.

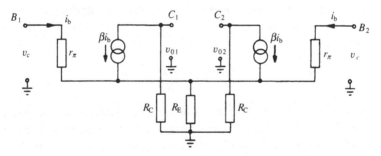

Fig. 2.4. Simplified equivalent circuit for common-mode voltage gain.

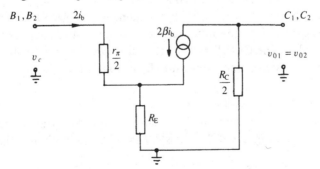

Thus the common-mode gain

$$A_c = \frac{v_{o1}}{v_c} = \frac{-\beta R_c}{r_\pi + 2(\beta+1)R_E} \qquad (2.6)$$

Since $g_m = \dfrac{\beta}{r_\pi}$ and $\beta \gg 1$

$$A_c \simeq \frac{-g_m R_c}{1 + 2g_m R_E} \qquad (2.7)$$

Also

$$g_m R_E \gg 1, \quad \therefore \; A_c \simeq -\frac{R_c}{2R_E} \qquad (2.8)$$

It should be noted that the common-mode gain is the same in magnitude and phase for both outputs, v_{o1} and v_{o2}.

Common-mode rejection ratio

The common-mode rejection ratio is a useful figure of merit for the performance of a differential amplifier. It is defined as the ratio of the differential gain to the common-mode gain and gives a qualitative measure of the ability of a circuit to respond to difference signals while being insensitive to common-mode (in-phase) signals such as electrical noise. The common-mode rejection ratio (CMRR) may be calculated from (2.3) and (2.8).

$$\mathrm{CMRR} = \frac{A_d}{A_c} \simeq -\frac{g_m R_c}{2} \bigg/ -\frac{R_c}{2R_E} \simeq g_m R_E \qquad (2.9)$$

It is apparent from (2.9) that a high value of the resistor R_E will give a large CMRR, but it should be remembered that any increase in resistance, for a fixed supply voltage will reduce the collector current I_c and hence the value of g_m.* Therefore, in order to obtain a large CMRR it is necessary to increase the value of R_E and, at the same time, to increase the supply voltage so that the collector current preferably remains unchanged. This may be achieved by connecting the emitter resistance to a negative supply voltage of a larger value. The CMRR may be further improved by substituting a constant current source for the resistor R_E in Fig. 2.1.

Worked example 2.1

What is the common-mode rejection ratio of the ideal differential amplifier shown in Fig. 2.1, if $R_c = 1.5\,\mathrm{k\Omega}$, $R_E = 1\,\mathrm{k\Omega}$ and both the collector

* $\quad g_m = \dfrac{q}{kT} I_c$ Siemens.

currents are 1.5 mA? If the input terminal B_2 is grounded and a sinusoidal signal of 3.5 mV RMS is applied to the input terminal B_1, then which output terminal would give a non-inverting gain? What would be the output voltage of the amplifier at that terminal? Assume that the circuit works in an environment having a temperature of 27 °C.

Solution

The transconductance of the transistors at a temperature of 27 °C is

$$g_m = \frac{q}{kT} I_c = \frac{1.6 \times 10^{-19} \times 1.5 \times 10^{-3}}{1.38 \times 10^{-23} \times 300} = 0.058 \text{ S}$$

Therefore, from (2.9),

$$\text{CMRR} = g_m R_E = 0.058 \times 1 \times 10^3 = 58$$

or

$$\text{CMRR}_{dB} = 20 \log_{10} \text{CMRR} = 20 \log_{10} 58 = 35.3 \text{ dB}$$

By observing (2.4) it can be said that the terminal C_2 gives non-inverting gain.

Again using (2.4) the output voltage at this terminal can be calculated to be

$$v_{o2} = \frac{g_m R_c}{2}(v_{i1} - v_{i2}) = \frac{0.058 \times 1.5 \times 10^3}{2}(3.5 \times 10^{-3} - 0)$$

$$= 0.15 \text{ volts}$$

Worked example 2.2

Transistors Q_1 and Q_2 of the circuit in Fig. ex.2.2 are matched and each has β equal to 85. Find (a) the d.c. output voltages V_{o1} and V_{o2}, (b) the single-ended differential voltage gain and (c) the double-ended differential voltage gain. If the collector output resistance of Q_3 is 300 kΩ, also find the common-mode rejection ratio. Assume that the circuit works in an environment at a temperature of 18 °C.

Solution

The voltage at the base of Q_3,

$$V_{B3} = -12 \cdot \frac{1.2 \times 10^3}{(1.2 + 3.6) \times 10^3} = -3 \text{ V}$$

Therefore the voltage at the emitter of Q_3,

$$V_{E3} = -3 - 0.7 = -3.7 \text{ V}$$

Fig. ex.2.2

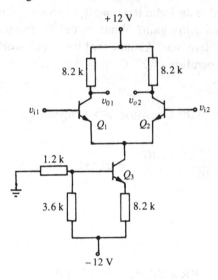

and the emitter current of Q_3,

$$I_{E3} = \frac{-(3.7-12)}{8.2 \times 10^3} = 1.012\,\text{mA}.$$

Hence

$$I_{C3} = \frac{\beta}{1+\beta} \cdot I_{E3} = \frac{85}{1+85} \times 1.012 \times 10^{-3}$$

$$= 1\,\text{mA}.$$

Again

$$I_{C1} = I_{C2} = \frac{I_{C3}}{2} - \frac{I_{C1}}{\beta}$$

or

$$I_{C1} = I_{C2} = \frac{\beta}{1+\beta} \cdot \frac{I_{E3}}{2} = \frac{85}{1+85} \cdot \frac{1.012 \times 10^{-3}}{2} = 0.5\,\text{mA}.$$

The d.c. collector voltages are, therefore,

$$V_{o1} = V_{o2} = 12 - 0.5 \times 10^{-3} \times 8.2 \times 10^3$$

$$= 7.9 \text{ volts}.$$

From (2.3) the single-ended differential gain,

$$A'_{d(s)} = -\frac{g_m R_c}{2} = \frac{-1.6 \times 10^{-19} \times 0.5 \times 10^{-3} \times 8.2 \times 10^3}{1.38 \times 10^{-23} \times 291 \times 2}$$

$$= -81.67$$

From (2.5) the double-ended differential gain,

$$A_{d(d)} = -g_m R_c = -163.34$$

The effective $R_E = (300 + 8.2) \times 10^3$

$$= 308.2 \text{ k}\Omega$$

Therefore from (2.9) the common-mode rejection ratio is

$$g_m R_E = \frac{1.6 \times 10^{-19} \times 0.5 \times 10^{-3}}{1.38 \times 10^{-23} \times 291} \times 308.2 \times 10^3$$

$$= 6139.6$$

or

$$\text{CMRR}_{dB} = 20 \log_{10} 6139.6 = 75.8 \text{ dB}$$

2.1.2 Constant current sources

Current sources are as important and as useful as voltage sources. They are widely used as emitter sources for differential amplifiers and also to bias transistors. The simplest approximation to a current source is a resistor in series with a voltage source. However, in order to obtain a closer approximation to a true constant current source from a finite supply voltage a transistor has to be used. Figure 2.5 shows some examples of

Fig. 2.5. Different types of current sources.

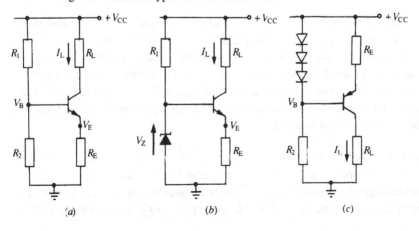

transistor current sources. The base potential can be provided in a number of ways using a voltage divider, a Zener diode or a few forward biased diodes in series, as shown in the diagram. In all cases a constant current is provided, but the range of load voltage depends on the biasing resistors, the Zener diode and the number of diodes. The base potential, V_B, of the circuits in Figs. 2.5(a) and (b), is fixed at a voltage very much greater than the base-emitter 'on' voltage of the transistor. By observing Figs. 2.5(a) and 2.5(b) we may write

$$V_E = V_B - V_{BE(ON)}$$

$$I_E = \frac{V_E}{R_E} = (V_B - V_{BE(ON)})/R_E \tag{2.10}$$

But, since $I_E \simeq I_C$ for large β

$$I_C \simeq (V_B - V_{BE(ON)})/R_E \tag{2.11}$$

So the collector current is independent of the collector voltage, as long as the transistor is not saturated, i.e. the collector voltage is greater than $V_E + V_{CE(sat)}$. Since $V_B \gg V_{BE}$ any variation in V_{BE} due to temperature change (typically $2\,\text{mV}/^\circ\text{C}$) will have negligible effect on the collector current.

In the circuit in Fig. 2.5(c) the base potential should be less than the emitter voltage V_E by the amount $V_{BE(ON)}$, so that the transistor keeps conducting. Now the collector current is given by

$$I_C \simeq I_E \simeq \frac{V_{CC} - V_E}{R_E} \simeq \frac{V_{CC} - [V_B + V_{BE(ON)}]}{R_E}$$

$$\simeq \frac{V_{CC} - [V_{CC} - nV_{BE(ON)} + V_{BE(ON)}]}{R_E} \simeq \frac{(n-1)V_{BE(ON)}}{R_E} \tag{2.12}$$

where n is the number of diode-connected transistors. Thus the collector current in this case is also independent of the collector voltage.

A voltage-programmable current source may be obtained by applying a varying voltage at the base of the transistor. The swing of this input signal must be small enough so that the emitter voltage never reaches zero potential. Then the output current of the current source will be proportional to the input voltage.

2.1.3   Current mirrors

Current mirrors are very popular in integrated circuits and are made using the matched base-emitter technique which works as follows. In the basic circuit, shown in Fig. 2.6, the transistor Q_2 acts as a diode with a

Fig. 2.6. Basic current mirror circuit.

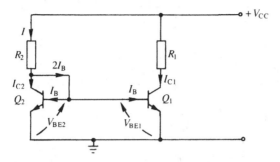

base-emitter voltage V_{BE} across it. The base-emitter voltage of transistor Q_1 is also equal to V_{BE}. Transistors Q_1 and Q_2 are identical and have similar variations of V_{BE} and β with temperature; and since their bases are at identical potentials, their collector currents will also be equal. If the current gain β of the transistors is very large, then we can neglect the base currents and write

$$I_{C1} = I_{C2} \simeq I = \frac{V_{CC} - V_{BE}}{R_2} \tag{2.13}$$

If the current gain β of each transistor is small, we cannot neglect the base currents, and in this case the collector currents

$$I_{C1} = I_{C2} = \frac{V_{CC} - V_{BE}}{R_2} - 2I_B$$

$$= \frac{V_{CC} - V_{BE}}{R_2} - \frac{2I_{C1}}{\beta}$$

or,

$$\beta I_{C1} + 2I_{C1} = \frac{(V_{CC} - V_{BE})\beta}{R_2}$$

or,

$$I_{C1} = (V_{CC} - V_{BE})\frac{\beta}{(\beta + 2)R_2} \tag{2.14}$$

From (2.14) it can be noted that the value of collector current I_{C1} is now less than that of the current I by an amount $\frac{2}{\beta + 2} \cdot I$.

Fig. 2.7. Modified current mirror circuit.

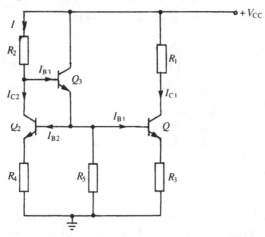

This can be compensated by deriving the base currents for transistors Q_1 and Q_2 through an emitter follower as shown in the modified circuit in Fig. 2.7. The emitter follower comprises the transistor Q_3 and the resistor R_5, and its collector current is of similar order to those of the base currents of the transistors Q_1 and Q_2. Resistors R_3 and R_4 have equal values. From the figure it can be noted

$$I_{C2} > I_{B2}$$

Again $I_{B2} > I_{B3}$. Therefore, $I_{C2} \gg I_{B3}$. Hence

$$I_{C1} = I_{C2} \simeq I \qquad (2.15)$$

When R_3 has a different value to R_4, the collector currents of the transistors Q_1 and Q_2 will be in the ratio of the two resistors.

In resistorless integrated circuit-operational amplifiers the operating current of the whole amplifier may be set by one external resistor only, which is R_2 in our case, with all the quiescent currents of the individual amplifier stages inside being controlled by current mirrors. For the simple circuit of Fig. 2.6 a suitable value of R_2 can be determined from the knowledge of V_{CC} and the current needed by the load which is R_1.

Current mirrors can be expanded to source current to several loads as shown in Fig. 2.8. If n matched transistors are used in the circuit, it can be shown that the collector currents

$$I_1, I_2, \ldots, I_{n-1} = I_C = I \cdot \frac{\beta}{\beta + n} \qquad (2.16)$$

Fig. 2.8. Multiple repeater circuit.

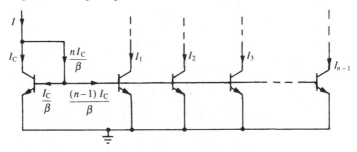

Worked example 2.3

The circuit of Fig. ex.2.3 is part of an integrated circuit. All the transistors in the circuit are identical each having a current gain of 110. Determine the value of R if the collector currents of Q_2, Q_3, Q_4 and Q_5 have to be equal to 2.9 mA. The base-emitter 'on' voltage of all the transistors is 0.7 volts.

Solution

Since all the transistors are identical

$$I_{C1} = I_{C2} = I_{C3} = I_{C4} = I_{C5} = 2.9 \times 10^{-3} \, \text{A}$$

From (2.16)

$$I = I_C \left(\frac{\beta + n}{\beta} \right) = 2.9 \times 10^{-3} \left(\frac{110 + 5}{110} \right) = 3.03 \, \text{mA}$$

$$\therefore R = \frac{9 - 0.7}{3.03 \times 10^{-3}} = 2.74 \, \text{k}\Omega.$$

Fig. ex.2.3

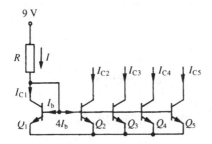

Fig. ex.2.4

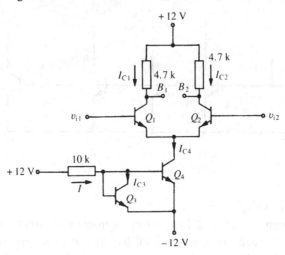

Worked example 2.4

Transistors Q_1, Q_2, Q_3 and Q_4 in the circuit of Fig. ex.2.4 are all identical and each has a current gain of 55. Find the d.c. output voltages at terminals B_1 and B_2. The base-emitter 'on' voltage of the transistors is 0.7 volts.

Solution

The transistors Q_3 and Q_4 form a current mirror. Therefore

$$I_{C4} = I_{C3}$$

Again

$$I_{C3} + 2\frac{I_{C3}}{\beta} = I$$

Thus

$$I_{C4} = I_{C3} = \frac{I\beta}{(2+\beta)}$$

$$= \frac{12-(-12)-0.7}{10 \times 10^3} \times \frac{55}{2+55} = 2.25\,\text{mA}$$

$$I_{E1} = I_{E2} = \frac{I_{C4}}{2} = \frac{2.25}{2} \times 10^{-3} = 1.125 \times 10^{-3}\,\text{A}$$

So

$$I_{C1} = \frac{\beta}{1+\beta} I_{E1} = 1.105 \, \text{mA}$$

$$\therefore \ V_{B1} = V_{B2} = 12 - 4.7 \times 10^3 \times 1.105 \times 10^{-3} = 6.8 \, \text{V}$$

2.1.4 Darlington connection

If we connect two transistors together as shown in Fig. 2.9, the configuration is called a Darlington pair. It behaves like a single transistor with an effective current gain approximately equal to the product of the current gains of the two transistors. The base-emitter 'on' voltage of the pair is twice the base-emitter 'on' voltage of a single transistor and the saturation voltage is greater than that of a single transistor by an amount equal to the base-emitter 'on' voltage. If a small signal current I_i is input to the base, the collector current of transistor Q_1 is $\beta_1 I_i$ and the emitter current is $(\beta_1 + 1)I_i$. The latter becomes the base current of transistor Q_2 and hence the collector current of Q_2 is $\beta_2(\beta_1 + 1)I_i$. Thus the total collector current of the Darlington circuit is

$$\beta_2(\beta_1 + 1)I_i + \beta_1 I_i \simeq \beta_1 \beta_2 I_i \tag{2.17}$$

Therefore this circuit can be very useful in high-current stages where a large gain is required, e.g. at the output stages of power amplifiers or in voltage regulators. For a typical power Darlington transistor the current gain can be 5000 at a collector current of 10 amperes.

Another interesting application of the Darlington circuit is as an emitter follower. It provides an excellent buffering between a high impedance source and a low impedance load.

Fig. 2.9. Darlington circuit.

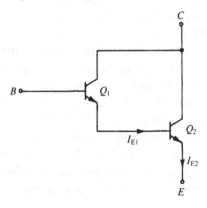

Small signal effective input and emitter resistance

The a.c. resistance looking into the base of Q_2 of the circuit in Fig. 2.9 is given by

$$r_{in\,2} \simeq \beta_2 r_{e2} \tag{2.18}$$

where r_{e2} is the small-signal base-emitter resistance of Q_2. We can say

$$r_{e2} \simeq \frac{kT}{qI_{E2}}$$

Since $r_e = \dfrac{1}{g_m}$ for small-signal models. Therefore the small-signal base-emitter resistance of Q_1,

$$r_{e1} \simeq \frac{kT}{qI_{E1}} = \frac{kT}{qI_{E2}/\beta_2} = \frac{\beta_2 kT}{qI_{E2}} = \beta_2 r_{e2} \tag{2.19}$$

From Fig. 2.9 we may observe that the effective input resistance of the Darlington pair is

$$r_{in} = \beta_1(r_{e1} + r_{in\,2}) \tag{2.20}$$

Substituting for r_{e1} and $r_{in\,2}$ in (2.20) from (2.19) and (2.18) respectively, we obtain,

$$r_{in} = \beta_1[\beta_2 r_{e2} + \beta_2 r_{e2}] = 2\beta_1 \beta_2 r_{e2} \tag{2.21}$$

The effective emitter resistance can now be obtained from (2.21)

$$r_e = \frac{r_{in}}{\beta_1 \beta_2} = \frac{2\beta_1 \beta_2 r_{e2}}{\beta_1 \beta_2} = 2r_{e2} \tag{2.22}$$

Darlington pairs are usually fabricated on a single chip and Q_1 and Q_2 have the same characteristics. Therefore we may assume $\beta_1 = \beta_2 = \beta$.

Darlington pairs are available commercially constructed on a single chip, as single packages, usually with a resistor connected across the base and emitter of Q_2 as shown in Fig. 2.10. Resistor R prevents leakage current through transistor Q_1 from biasing Q_2 into conduction. Its value is so chosen that when the combination is off, the leakage current of transistor Q_1 produces a voltage drop across resistor R which is smaller than the base-emitter 'on' voltage of transistor Q_2; again when the combination is on, the current through resistor R is small compared to the base current of transistor Q_2. The leakage current in small-signal transistors is of the order of nanoamperes and in power transistors it is as much as hundreds of

Fig. 2.10. Improved Darlington circuit.

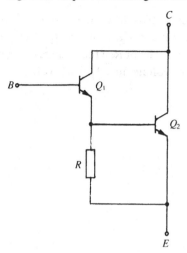

microamperes. Therefore for a small-signal transistor Darlington R might have a value of a few thousand ohms and for a power transistor Darlington a few hundred ohms.

Worked example 2.5

A Darlington pair draws a collector current of 2 mA. If the current gains of the two transistors are 75 and 100, find the total current gain and the effective input and emitter resistances at a room temperature of 27 °C.

Solution

The total current gain $\beta = \beta_1\beta_2 = 75 \times 100 = 7500$. From (2.21) the effective input resistance

$$r_{in} = 2\beta_1\beta_2 r_{e2} \simeq 2\beta_1\beta_2\frac{kT}{qI_{E2}} = 2\beta_1\beta_2\frac{kT}{q(I_{C2}+I_{E1})}$$

$$\simeq 2\beta_1\beta_2\frac{kT}{qI_C}$$

$$= \frac{2 \times 75 \times 100 \times 1.38 \times 10^{-23} \times 300}{1.6 \times 10^{-19} \times 2 \times 10^{-3}} = 194\,k\Omega$$

From (2.22) the effective emitter resistance

$$r_e \simeq \frac{r_{in}}{\beta_1\beta_2} = \frac{194\,000}{7500} = 25.8\,\Omega$$

Worked example 2.6

The quiescent collector current is 10A for the power transistor Darlington shown in Fig. ex.2.6. The gains of transistors Q_1 and Q_2 are 200 and 100 respectively. Find a suitable value for R, if the leakage current of Q_1 is 0.2 mA and the base-emitter 'on' voltage for Q_2 is 0.7 volts.

Fig. ex.2.6

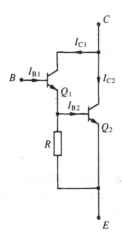

Solution

When the Darlington pair is OFF,

$$I_{\text{leakage}} \times R < V_{\text{BE(ON)}}$$

or

$$R < \frac{V_{\text{BE(ON)}}}{I_{\text{leakage}}} = \frac{0.7}{0.2 \times 10^{-3}} = 3.5 \,\text{k}\Omega$$

When the Darlington pair is ON, the current through R,

$$I_R \ll I_{B2}$$

So we may neglect I_R and have

$$I_{C1} + I_{C2} = 10 \tag{1}$$

$$I_{C1} + I_{B1} = I_{B2} \tag{2}$$

$$I_{C1} = 200 \, I_{B1} \tag{3}$$

and

$$I_{C2} = 100 \, I_{B2} \tag{4}$$

From (2) and (3)

$$201\, I_{B1} = I_{B2} \tag{5}$$

and from (1), (3), (4) and (5)

$$200\, I_{B1} + 100 \times 201 \times I_{B1} = 10$$

or

$$I_{B1} = \frac{10}{20.3}\,\text{mA}$$

from (3)

$$I_{C1} = \frac{2000}{20.3}\,\text{mA}$$

Therefore from (2)

$$I_{B2} = \frac{2010}{20.3} = 99\,\text{mA}$$

Again when the Darlington pair is ON,

$$\frac{V_{BE(ON)}}{R} \ll I_{B2}$$

Hence

$$R \gg \frac{V_{BE(ON)}}{I_{B2}} = \frac{0.7}{99 \times 10^{-3}} = 7\,\Omega.$$

2.1.5   Level shifting circuits

In integrated circuits the use of coupling capacitors is always avoided and to offset any direct voltage level present between say, two amplifier stages, other methods are employed. One of these methods is to use a potential divider connected between the output and the supply rail as shown in Fig. 2.11. Resistors R_1 and R_2 lower the direct voltage level of the

Fig. 2.11. Basic level shifting circuit.

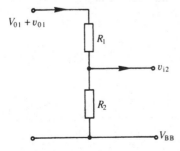

output signal but, at the same time attenuate the a.c. signal. If we assume that the output of the first stage consists of a signal v_{o1} superimposed on a direct voltage V_{o1} and choose resistors R_1 and R_2 such that the input to the second stage has no direct voltage superimposed on the a.c. signal v_{i2}, then we may write

$$\left(\frac{V_{o1} - V_{BB}}{R_1 + R_2}\right) R_1 = V_{o1}$$

or

$$R_2/R_1 = -V_{BB}/V_{o1} \tag{2.23}$$

Again,

$$\frac{v_{i2}}{v_{o1}} = \frac{R_2}{R_1 + R_2} = -\frac{V_{BB}}{V_{o1} - V_{BB}} \tag{2.24}$$

This shows that the attenuation of the signal v_{o1} can be reduced by making $V_{BB} \gg V_{o1}$. In practice, however, the magnitude of V_{BB} is limited and a considerable attenuation occurs.

In order to keep the attenuation of input signals as low as possible we use two transistors in the level shifting circuit as shown in Fig. 2.12. Transistor Q_1 operates as an emitter follower, whereas, transistor Q_2 in series with the resistor R_2 forms a constant current source since the base of transistor Q_2 is held at a constant voltage, V_B, by connecting transistor Q_3 as a diode. The voltage drop across the resistor R_1 removes any offset direct voltage which would have existed at the input of the second stage otherwise. The output resistance of the current source is very large and hence the attenuation of

Fig. 2.12. Level shifting circuit with transistors.

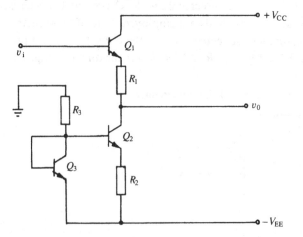

Fig. 2.13. Level shifting circuit with diodes.

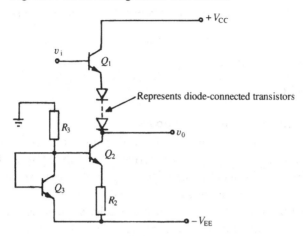

Represents diode-connected transistors

the output signal of the first stage is very small. In this circuit the base-emitter voltage of Q_3 is equal to that of Q_2 together with the voltage drop across R_2. The difference in the base-emitter voltage of Q_2 and Q_3 defines the collector current of Q_2.

Another technique of introducing a direct voltage level shift is to replace the resistor R_1 by n forward-biased diode-connected transistors as shown in Fig. 2.13. This will produce a level shift of approximately $n \times 0.7$ V.

Worked example 2.7

The input signal v_1 to the circuit of Fig. ex.2.7 has a d.c. level of 3.5 volts. What will be the value of R_1 so that the mean voltage at the output is zero?

Fig. ex.2.7

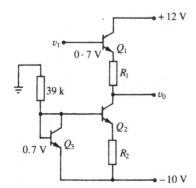

Solution

Transistors Q_2 and Q_3 form a current mirror. The d.c. current I is

$$0-[-10-(-0.7)]/39 \times 10^3 = 0.24\,\text{mA}$$

$$\therefore \qquad R_1 = (3.5-0.7)/0.24 \times 10^{-3}$$

$$= 11.67\,\text{k}\Omega$$

2.2 Cascaded amplifier stages

In many applications, an amplifier with a single transistor cannot provide all the gain that is required to drive a particular type of load. Several amplifier stages may be needed to amplify the input signal to a sufficient level. The output of one stage is normally connected to the input of another, and the stages are said to be in cascade. Sometimes amplifier stages of different configurations are cascaded to provide source and load impedances of a required magnitude.

To analyse an amplifier circuit with two stages in cascade, we may make use of the voltage gain A_v, the current gain A_i, the input impedance Z_i and the output admittance Y_o of each stage in terms of the hybrid parameters of the transistor used in that stage.

The hybrid parameters of a two-port device can be derived by considering the diagram shown in Fig. 2.14. The terminal behaviour of the device can be described by the two voltages and two currents. If we choose the current i_1 and the voltage v_1 as the independent variables and assume that the device is linear, we may write,

$$v_1 = h_i i_1 + h_r v_2$$
$$i_2 = h_f i_1 + h_o v_2$$

where

h_i is the input resistance with the output short circuited
h_r is the reverse open-circuit voltage gain
h_f is the short-circuit current gain

and

h_o is the output conductance with the input open circuited.

Fig. 2.14. Two-port devices.

Fig. 2.15. (a) Common-emitter, (b) Common-base and (c) Common-collector configurations.

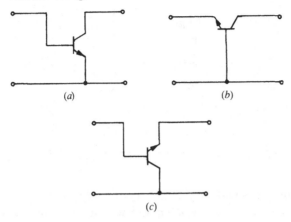

(a)   (b)

(c)

Since transistors can be connected in various configurations as two-port devices (common-emitter, common-base or common-collector, see Fig. 2.15), and the values of the hybrid parameters vary from one configuration to another, it is convenient to add another subscript (b, e or c) to the above notations of the hybrid parameters to designate the type of configuration. These parameter values are usually obtained from the manufacturers' data sheets.

For small-signal analysis of a single stage transistor amplifier, the voltage gain A_v, the current gain A_I, the input impedance, Z_i and the output admittance, Y_o, are respectively given by

$$A_v = A_I \frac{Z_L}{Z_i}, \quad A_I = \frac{-h_f}{1 + h_o Z_L}$$

$$Z_i = h_i + h_r A_I Z_L, \quad Y_o = h_o - \frac{h_f h_r}{h_i + R_s} \tag{2.25}$$

where Z_L is the load impedance and R_s is the source impedance.

Taking into account the source impedance, the current amplification is given by

$$A_{IS} = A_I \frac{R_s}{Z_i + R_s} \tag{2.26}$$

and the voltage amplification is given by

$$A_{VS} = A_V \frac{Z_i}{Z_i + R_s} \tag{2.27}$$

Fig. 2.16. Multistage amplifier.

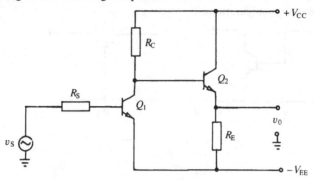

Let us now consider a two stage amplifier circuit shown in Fig. 2.16. The first stage is connected in a common-emitter configuration and the second stage is connected in common-collector configuration. From (2.25) we may note that the current gains have to be found before the input impedances or the voltage gains of the stages can be calculated. Again since the load impedance of the first stage can not be computed until we know the input impedance of the second stage, it is logical that we start analysing the circuit with the second stage.

Input impedance of cascaded amplifiers

The small-signal equivalent circuit of the cascaded amplifier of Fig. 2.16 is shown in Fig. 2.17. Since the second stage is connected in common-collector configuration, the current gain, the input impedance and the voltage gain of this stage may be written in the following forms using (2.25),

$$A_{12} = -\frac{I_{e2}}{I_{b2}} = -\frac{h_{fc}}{1 + h_{oc}R_E} \tag{2.28}$$

$$R_{i2} = h_{ic} + h_{rc}A_{12}R_E \tag{2.29}$$

Fig. 2.17. Small-signal circuit of the multistage amplifier.

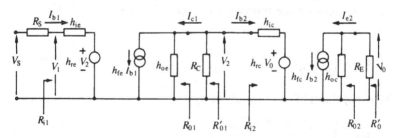

and

$$A_{V2} = \frac{V_o}{V_2} = A_{12}\frac{R_E}{R_{i2}} \tag{2.30}$$

Now let us consider the first stage. The load impedance R_{L1} of this stage is the parallel combination of the collector resistance R_C and the input impedance of the second stage R_{i2}. Thus

$$R_{L1} = \frac{R_C \cdot R_{i2}}{R_C + R_{i2}} \tag{2.31}$$

Now the current gain of the first stage is

$$A_{I1} = -\frac{I_{C1}}{I_{b1}} = \frac{-h_{fe}}{1 + h_{oe}R_{L1}} \tag{2.32}$$

and the input impedance of the first stage, which is the input impedance of the amplifier, is

$$R_{i1} = h_{ie} + h_{re}A_{I1}R_{L1} \tag{2.33}$$

Output impedance of cascaded amplifiers
The voltage gain of the first stage is given by

$$A_{V1} = A_{I1}\frac{R_{L1}}{R_{i1}} \tag{2.34}$$

and the output impedance of transistor Q_1 is

$$R_{o1} = 1/Y_{o1} = \frac{1}{h_{oe} - \dfrac{h_{fe}h_{re}}{h_{ie} + R_s}} \tag{2.35}$$

Since this impedance is in parallel with the resistor R_C, the output impedance of the first stage is

$$R'_{o1} = \frac{R_c R_{o1}}{R_c + R_{o1}} \tag{2.36}$$

Now the effective source impedance of transistor Q_2 is $R'_{s2} = R'_{o1}$ and its output impedance,

$$R_{o2} = 1/Y_{o2} = \frac{1}{h_{oc} - \dfrac{h_{fc}h_{rc}}{h_{ic} + R'_{s2}}} \tag{2.37}$$

The output impedance of the amplifier is, therefore, given by

$$R'_\text{o} = \frac{R_{\text{o}2} R_\text{E}}{R_{\text{o}2} + R_\text{E}}$$ (2.38)

Overall current gain

The overall current gain is the ratio of the emitter current of the second stage to the base current of the first stage and from Fig. 2.17 we have

$$A_\text{I} = -\frac{I_{\text{e}2}}{I_{\text{b}1}} = -\frac{I_{\text{e}2}}{I_{\text{b}2}} \cdot \frac{I_{\text{c}1}}{I_{\text{b}1}} \cdot \frac{I_{\text{b}2}}{I_{\text{c}1}}$$

$$= -A_{\text{I}2} \cdot A_{\text{I}1} \cdot \frac{I_{\text{b}2}}{I_{\text{c}1}} = A_{\text{I}2} \cdot A_{\text{I}1} \frac{R_\text{c}}{R_\text{c} + R_{\text{i}2}}$$ (2.39)*

Overall voltage gain

The overall voltage gain is simply obtained by multiplying the individual voltage gain of each stage of the amplifier.

$$A_\text{v} = A_{\text{v}2} A_{\text{v}1}$$ (2.40)

If we take the source impedance into account, the current gain becomes

$$A_{\text{IS}} = A_\text{I} \cdot \frac{R_\text{S}}{R_{\text{i}1} + R_\text{S}}$$ (2.41)

and the voltage gain of the amplifier becomes

$$A_{\text{vs}} = V_\text{o}/V_\text{s} = A_\text{v} \cdot \frac{R_{\text{i}1}}{R_{\text{i}1} + R_\text{S}}$$ (2.42)

It was mentioned earlier that the values of hybrid parameters of a specific transistor vary from configuration to configuration, see Table A1 in Appendix A. So we can say by examining (2.25) that different transistor amplifier configurations have different values of voltage gain, current gain, input impedance and output impedance. Assuming $R_\text{s} = 1\,\text{k}\Omega$ and $R_\text{L} = 4.7\,\text{k}\Omega$, and using Table A1 we may find the values of various quantities which may be used to choose the right transistor configuration for a particular stage in multistage amplifier circuits. These are shown in Table 2.1.

The factors governing the choice of transistor configuration for different stages within an amplifier are given below but we must remember that they are only guidelines and the choice may vary from one circuit to another.

---

* $(I_{\text{c}1} + I_{\text{b}2}) R_\text{c} = -I_{\text{b}2} \cdot R_{\text{i}2}$

Table 2.1. *Input and output impedances and current and voltage gains of various transistor amplifier configurations*

Quantity	Common-collector	Common-base	Common-emitter
A_I	46	0.98	-45
A_V	0.99	201	-201
R_i	$215\,\text{k}\Omega$	$23\,\Omega$	$1047\,\Omega$
R_o	$41\,\Omega$	$1.3\,\text{M}\Omega$	$55\,\text{k}\Omega$

Input Stage. The frequency response of the source whose signal is to be amplified may depend upon the impedance into which it operates. Some of the sources need essentially open-circuit or short-circuit operation. In such cases the common-collector or common-base configuration is used although they may not provide the necessary voltage or current gain. From Table 2.1 we may notice that the common-collector configuration has a high input impedance, whereas, the common-base configuration has a low input impedance. The former circuit has, in addition, a low output impedance and is widely used as a buffer stage between a high-impedance source and low-impedance load. The common-base configuration is used to match a very low-impedance source and also is used as a constant-current source in many applications.

Intermediate Stage. Table 2.1 also shows that the voltage gain of a common-collector configuration is less than unity. Hence we may not increase the overall voltage amplification by cascading such configurations. Again the voltage gain of two or more common-base circuits in cascade is about the same as that of the last stage alone. This can be verified as shown below,

$$h_{fb} < 1$$

So

$$A_I = -\frac{-h_{fb}}{1 + h_{ob}R_L} < 1 \qquad (2.43)$$

Again, the voltage gain of the first stage is given by

$$A_V = A_I \frac{R_L}{R_i} = A_I \frac{R_C // R_{i2}}{R_{i1}} = A_I \frac{R_C R_{i2}}{(R_C + R_{i2})R_{i1}} \qquad (2.44)$$

Now if we make the first and second stages identical, then $R_{i1} = R_{i2}$ and from (2.44) we may write

$$A_V < A_I$$

For a single stage the current gain is less than unity, so

$$A_V < 1$$

Therefore this configuration is seldom used in cascaded form.

The only remaining configuration is the common-emitter type and since its current gain is much greater than unity, a large voltage amplification can be achieved by cascading such stages. So, in the intermediate stages of an amplifier, where a large voltage gain is a requirement, transistors may be connected in the common-emitter configuration.

Output Stage. In many applications an amplifier is required to drive a low impedance load. In such applications, by examining the Table 2.1, we may say that the common-collector configuration is the one which could be used for the output stage, because it has a very low output impedance.

Frequency response

In practice, the individual stages of a multistage amplifier may have some lower cutoff frequencies that are equal in magnitude, and others which are not. The same applies to the upper cutoff frequencies. For the special cases where all stages of a multistage amplifier have identical lower cutoff frequencies or identical upper cutoff frequencies, the overall cutoff frequencies may be obtained as follows.

Lower cutoff frequency, $f_{L(overall)}$. The voltage gain of a single stage amplifier is given by

$$A_L = \frac{1}{\sqrt{[1 + (f_L/f)^2]}}$$

where f_L is the lower cutoff frequency.

Therefore, the gain of an amplifier with n identical stages,

$$A_L{}^n = \left(\frac{1}{\sqrt{[1 + (f_L/f)^2]}} \right)^n = \frac{1}{\sqrt{\left[1 + \left(\frac{f_{L(overall)}}{f} \right)^2 \right]}} \tag{2.45}$$

where $f_{L(overall)}$ is the overall lower cutoff frequency. Thus in the lower frequency range the overall gain of the amplifier drops by a factor of $1/\sqrt{2}$ (3 dB) at $f = f_{L(overall)}$.*

Upper cutoff frequency, $f_{u(overall)}$. The voltage gain of a single stage amplifier is also given by

$$A_U = \frac{1}{\sqrt{[1 + (f/f_U)^2]}}$$

where f_U is the upper cutoff frequency.

* $f_{L(overall)} = f_L / \sqrt{(2^{1/n} - 1)}$.

Fig. 2.18. Frequency response of a multistage amplifier.

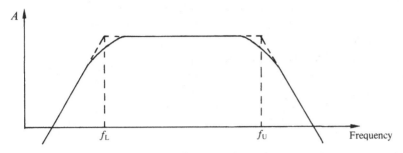

In a similar fashion as above, the gain of an amplifier with n identical stages will be

$$A_U{}^n = \left(\frac{1}{\sqrt{[1 + (f/f_U)^2]}}\right)^n = \frac{1}{\sqrt{[1 + (f/f_{U(overall)})^2]}} \qquad (2.46)$$

where $f_{U(overall)}$ is the overall upper cutoff frequency. Therefore in the upper frequency range the overall gain becomes $1/\sqrt{2}$ at $f = f_{U(overall)}$.†

By examining (2.45) and (2.46), we may say that the overall lower cutoff frequency becomes larger as the number of individual stages is increased. At the same time the overall upper cutoff frequency becomes smaller. Hence the bandwidth becomes narrower. The overall frequency response falls off along asymptotes having slopes $20n$ dB/decade at frequencies outside the midband range (Fig. 2.18). If the lower cutoff frequencies of the individual stages are not similar in magnitude, the overall lower cutoff frequency is approximately equal to the largest of the stage lower cutoff frequencies. Again if the upper cutoff frequencies of the individual stages are not similar in magnitude the overall upper cutoff frequency is approximately equal to the smallest of the stage upper cutoff frequencies. In these cases calculation of the overall cutoff frequencies is very difficult and should be found out experimentally or with the help of a computer.

Worked example 2.8

In the two-stage amplifier shown in Fig. 2.16, the transistor parameters at the corresponding quiescent points are

$$h_{ie} = 1.4\,k\Omega, \; h_{fe} = 100, \; h_{re} = 2 \times 10^{-4},$$

$$h_{oe} = 2 \times 10^{-5}\,S, \; h_{ic} = 1.4\,k\Omega, \; h_{fc} = -101,$$

$$h_{rc} = 1 \quad \text{and} \quad h_{oc} = 2 \times 10^{-5}\,S$$

† $f_{U(overall)} = f_U[\sqrt{(2^{1/n} - 1)}]$.

Find the input and output impedances of the circuit and also the overall current and voltage gains. Given, $R_S = 1\,\text{k}\Omega$, $R_E = R_C = 4.7\,\text{k}\Omega$.

Solution

From (2.28) the current gain of the second stage

$$A_{12} = -\frac{h_{fc}}{1 + h_{oc}R_E} = -\frac{-101}{1 + 2 \times 10^{-5} \times 4.7 \times 10^3}$$

$$= 92.3$$

Hence from (2.29) the input resistance of the second stage

$$R_{i2} = h_{ic} + h_{rc}A_{12}R_E = 1.4 \times 10^3 + 1 \times 92.3 \times 4.7 \times 10^3$$

$$= 435.2\,\text{k}\Omega$$

and from (2.30) the voltage gain of the second stage

$$A_{V2} = A_{12} \cdot \frac{R_E}{R_{i2}} = 92.3 \times \frac{4.7 \times 10^3}{435.2 \times 10^3}$$

$$= 0.997$$

From (2.31) the load resistance of the first stage

$$R_{L1} = \frac{R_c R_{i2}}{R_c + R_{i2}} = \frac{4.7 \times 10^3 \times 435.2 \times 10^3}{(4.7 + 435.2)10^3}$$

$$= 4.65\,\text{k}\Omega$$

From (2.32) the current gain of the first stage

$$A_{11} = -\frac{h_{fe}}{1 + h_{oe}R_{L1}} = \frac{-100}{1 + 2 \times 10^{-5} \times 4.65 \times 10^3}$$

$$= -91.49$$

From (2.33) the input resistance of the first stage

$$R_{i1} = h_{ie} + h_{re}A_{11}R_{L1} = 1.4 \times 10^3 + 2 \times 10^{-4} \times (-91.49) \times 4.65 \times 10^3$$

$$= 1.315\,\text{k}\Omega$$

Hence the voltage gain of the first stage

$$A_{V1} = A_{11} \cdot \frac{R_{L1}}{R_{i1}} = -91.49 \times \frac{4.65 \times 10^3}{1.315 \times 10^3}$$

$$= -323.5$$

From (2.35) the output resistance of Q_1

$$R_{o1} = \cfrac{1}{h_{oe} - \cfrac{h_{fe}h_{re}}{h_{ie} + R_s}}$$

$$= \cfrac{1}{2 \times 10^{-5} - \cfrac{100 \times 2 \times 10^{-4}}{1.4 \times 10^3 + 1 \times 10^3}} = 85.71\,\text{k}\Omega$$

Therefore the output resistance of the first stage

$$R'_{o1} = \frac{R_c R_{o1}}{R_c + R_{o1}} = \frac{4.7 \times 10^3 \times 85.71 \times 10^3}{4.7 \times 10^3 + 85.71 \times 10^3}$$

$$= 4.45\,\text{k}\Omega$$

Again the output resistance of Q_2, from (2.37)

$$R_{o2} = \cfrac{1}{h_{oc} - \cfrac{h_{fc}h_{rc}}{h_{ic} + R'_{o1}}} = \cfrac{1}{2 \times 10^{-5} - \cfrac{-101 \times 1}{1.4 \times 10^3 + 4.45 \times 10^3}}$$

$$= 57.9\,\Omega$$

Hence the output resistance of the amplifier

$$R'_o = \frac{R_{o2}R_E}{R_{o2} + R_E} = \frac{57.9 \times 4.7 \times 10^3}{57.9 + 4.7 \times 10^3}$$

$$= 57\,\Omega$$

From (2.39) the overall current gain

$$A_i = A_{i2} \cdot A_{i1} \cdot \frac{R_c}{R_c + R_{i2}}$$

$$= -91.49 \times 92.3 \times \frac{4.7 \times 10^3}{4.7 \times 10^3 + 435.2 \times 10^3}$$

$$= -90.2$$

From (2.40) the overall voltage gain

$$A_V = A_{V2} \cdot A_{V1} = 0.997 \times -323.5$$

$$= -322.5$$

2.3   Tuned amplifiers

In many applications an amplifier is required to amplify a narrow band of frequencies centred about a certain frequency. Any signal containing only frequencies outside this band must not be amplified. Tuned amplifiers serve this purpose and are obviously very important in fields that depend on the simultaneous transmission of several channels of information. In radio broadcasting, several stations may simultaneously transmit signals in the same general geographic area and each station then uses a different carrier frequency. A radio receiver selects only one of these signals and this particular ability of the receiver is the key factor in radio communication. The same general idea is used in television receivers, two-way radio communications, long-distance telephone systems and also in the field of satellite communications. A tuned amplifier circuit is shown in Fig. 2.19. At resonance, the impedance of the tuned circuit may range from several kilohms up to a few megohms, depending on the components used.

Practical inductors can be represented by a resistance in series with a pure inductor and therefore, the tuned circuit of the amplifier shown in Fig. 2.19 can be redrawn as shown in Fig. 2.20. The Q of the circuit, which compares the energy stored to the energy dissipated per cycle, is equal to $\omega L/R$. When the circuit is at resonance

$$Q_0 = \frac{\omega_0 L}{R} \qquad\qquad (2.47)$$

where ω_0 is the resonant frequency and is equal to $1/\sqrt{(LC)}$.

Fig. 2.19. Tuned amplifier.

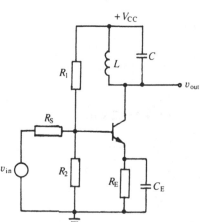

Fig. 2.20. Tuned circuit.

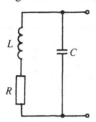

The impedance of the tuned circuit of Fig. 2.20 is given by

$$Z = \frac{1}{j\omega C} \left/\!\right/ (R + j\omega L) = \frac{R + j\omega L}{-\omega^2 CL + j\omega CR + 1} \qquad (2.48)$$

Using (2.48) we may plot the magnitude of the impedance of the tuned circuit as a function of frequency for various values of Q as shown in Fig. 2.21. It is obvious from the plot that if a constant current enters the tuned circuit at resonant frequency then a large voltage will develop across the circuit. At other frequencies the developed voltage will be considerably smaller.

Again for an inductor with a high value of Q the series RL circuit representing the inductor is equivalent to a parallel RL circuit over the frequency band of interest as shown in Fig. 2.22. It can be verified by finding the admittance of the series circuit

$$Y = \frac{1}{R + j\omega L} = \frac{R - j\omega L}{R^2 + \omega^2 L^2} = \frac{R}{\omega^2 L^2} + \frac{1}{j\omega L} \qquad (2.49)$$

and the admittance of the parallel circuit

$$Y_P = \frac{1}{R_P} + \frac{1}{j\omega L} \qquad (2.50)$$

Fig. 2.21. Magnitude of impedance as a function of frequency.

Fig. 2.22. Coil of the tuned circuit.

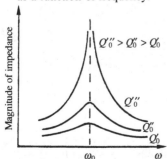

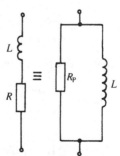

Fig. 2.23. Equivalent circuit of the tuned amplifier.

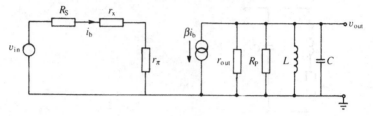

Comparing (2.49) and (2.50) we obtain

$$R_P = \frac{\omega^2 L^2}{R} = RQ^2 = \omega L Q^* \tag{2.51}$$

We may rewrite (2.51) for $\omega = \omega_0$ as follows,

$$Q_0 = R_P / \omega_0 L \tag{2.52}$$

Now we may draw the equivalent circuit of the tuned amplifier as shown in Fig. 2.23. The equivalent resistance of the resonant circuit is given by

$$R_{eq} = R_P // r_{out} \tag{2.53}$$

where r_{out} is the output resistance of the transistor. The effective Q of the circuit will now be

$$Q_{eff} = \frac{R_{eq}}{R_P} Q_0 \tag{2.54}$$

An expression for the voltage gain of the amplifier can be found by observing the equivalent circuit

$$A = \frac{v_{out}}{v_{in}} = \frac{-\beta i_b Z_L}{i_b (r_x + r_\pi + R_S)} = \frac{-\beta Z_L}{r_x + r_\pi + R_S} \tag{2.55}$$

where Z_L is the collector load impedance and the magnitude of the gain will be

$$|A| = \frac{\beta |Z_L|}{r_x + r_\pi + R_S} \tag{2.56}$$

At resonance, the peak gain occurs and it is given by

$$A_{res} = -\frac{\beta R_{eq}}{r_x + r_\pi + R_S} \tag{2.57}$$

* Usually Q is very high for tuned circuits: $Q = \dfrac{\omega L}{R} \gg 1.0$.

It can be shown with the help of (2.56) and (2.57) that the bandwidth between 3-dB frequencies for a high Q circuit is

$$f_U - f_L = BW = f_o/Q_{eff} \qquad (2.58)$$

where f_o is the resonant frequency, f_L is the lower cutoff frequency and f_U is the upper cutoff frequency. We may observe from (2.58) that a higher Q leads to a narrower bandwidth.

Worked example 2.9
In the circuit of Fig. 2.19 the resonant circuit has a Q of 100 and an inductance of $60\,\mu H$. The output resistance of the transistor is $40\,k\Omega$. If $r_x = 200\,\Omega$, $r_\pi = 2.5\,k\Omega$, $R_s = 3\,k\Omega$, $\beta = 75$ and $f_o = 1\,MHz$, find (a) the effective Q, (b) the gain at resonance and (c) the bandwidth of the amplifier.

Solution
From (2.52)

$$R_P = Q\omega_0 L = 100 \times 2\pi \times 1 \times 10^6 \times 60 \times 10^{-6} = 37.68\,k\Omega$$

From (2.53)

$$R_{eq} = R_P // r_{out} = \frac{37.68 \times 40 \times 10^6}{(37.68 + 40) \times 10^3} = 19.4\,k\Omega$$

(a)   From (2.54)

$$Q_{eff} = \frac{R_{eq}}{R_P} Q = \frac{19.4 \times 10^3}{37.68 \times 10^3} \times 100 = 51.5$$

(b)   From (2.57)

$$A_{res} = \frac{-\beta R_{eq}}{r_x + r_\pi + R_S} = \frac{-75 \times 19.4 \times 10^3}{200 + 2500 + 3000} = -255.3$$

(c)   From (2.58)

$$BW = \frac{f_o}{Q_{eff}} = \frac{1 \times 10^6}{51.5} = 19.42\,kHz$$

2.4   Power amplifiers

A power amplifier is often the last, or output, stage of an amplifying system and is designed to deliver a large amount of power to a load, whereas, the first few stages of an amplifying system, as mentioned in Section 2.2, may be designed to provide voltage amplification, or to provide buffering to a high impedance signal source. Power amplifiers are widely used in public address systems, high-fidelity systems, and radio and television receivers. In these applications, the load is usually a loudspeaker. Power amplifiers are also used in electromechanical control systems to drive electric servo motors.

There are several types of power amplifiers. The class A type conducts load current continuously during the complete cycle of a periodic input signal, and the output signal developed is always proportional to the input signal. A class B amplifier uses two transistors to drive a load. One transistor amplifies positive signal variations and the other amplifies negative signal variations, and the amplifier output is the combination of the waveform obtained from these two transistors. A class C power amplifier is one that conducts load current during less than one-half cycle of an input sine wave. It is extensively used in high frequency applications, such as radio-frequency transmitters.

2.4.1   Class A amplifier

A common-emitter amplifier that supplies power to a resistive load R_L is shown in Fig. 2.24. In the figure i_C, i_B and v_C represent the total instantaneous collector current, the total instantaneous base current and the total instantaneous collector-to-emitter voltage respectively. The quiescent values of the collector current, the base current and the collector-to-emitter voltage are given by I_C, I_B and V_C respectively. Again the instantaneous variation of the corresponding quantities from the quiescent values are represented by i_{CQ}, i_{BQ} and v_{CQ} respectively.

The static output characteristics of the amplifier and the current and voltage waveforms are shown in Fig. 2.25. The load line for the circuit passes through the point $i_C = 0$, $v_C = V_{CC}$, and $i_C = V_{CC}/R_L$, $v_C = 0$. That is, it

Fig. 2.24. Class A common-emitter amplifier with a resistive load.

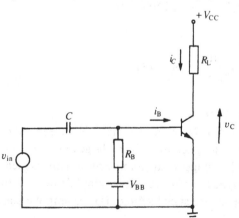

Fig. 2.25. The output characteristics and waveforms of the class A amplifier.

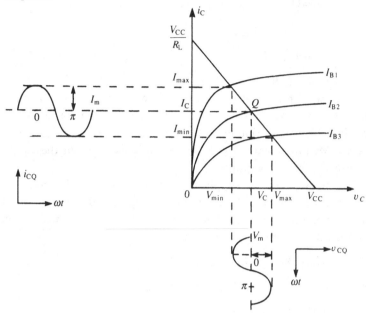

intersects the voltage axis at V_{CC} and the current axis at V_{CC}/R_L. The slope of the load line is $-1/R_L$ and it intersects the static curve at the quiescent operating point, Q which gives the current that will flow in the resistive load. Since we have assumed that the characteristics are equidistant for equal increments of input current i_{BQ}, the output current and hence the output voltage will be sinusoidal when the input current waveform is a sinusoid.

From the figure the output power is

$$P = \frac{V_m}{\sqrt{2}} \cdot \frac{I_m}{\sqrt{2}} = V_m I_m / 2 \qquad (2.59)$$

where I_m and V_m are the peak output current and voltage respectively. By examining Fig. 2.25, we may write

$$I_m = (I_{max} - I_{min})/2$$

and also

$$V_m = (V_{max} - V_{min})/2 \qquad (2.60)$$

$$I_{max} - I_{min} = (V_{max} - V_{min})/R_L$$

where I_{max}, I_{min}, V_{max} and V_{min} are the maximum and minimum values of output current and voltage swing respectively.

Now (2.59) becomes

$$P = (V_{max} - V_{min})(I_{max} - I_{min})/8$$

$$= (V_{max} - V_{min})^2/8R_L$$

$$= V_m^2/2R_L \qquad (2.61)$$

Similarly $P = I_m^2 R_L/2$.

Thus we can calculate the output power of the amplifier by plotting the load line on the volt–ampere characteristic of the transistor and then reading off the maximum and minimum values of voltage and current swing. Maximum output swing can be achieved by setting the Q point at the centre of the a.c. load line.

The average power supplied to the circuit is $V_{cc}I_C$; and the power absorbed by the output circuit is $I_C^2 R_L + I_m V_m/2$. Let us assume that the average power dissipated by the transistor is P_D. Then using the principle of conservation of energy we may write,

$$P_D = V_{cc}I_C - I_C^2 R_L - V_m I_m/2 \qquad (2.62)$$

but, since $V_{cc} = V_c + I_c R_L$, (2.62) yields,

$$P_D = V_c I_C - V_m I_m/2 \qquad (2.63)$$

By examining (2.63), we may say that the minimum dissipation occurs when $V_m I_m/2 = V_{cc}^2/8R_L$, since the maximum values which V_m and I_m can have, are $V_{cc}/2$ and $V_{cc}/2R_L (= I_C)$ respectively, as may be observed in Fig. 2.25. Again according to the same equation, the maximum dissipation occurs when the a.c. output power is zero, i.e., in the absence of any input signal. Then its magnitude is given by $V_c I_C$.

The collector efficiency is the ability of a transistor amplifier to convert the d.c. power of the supply into the a.c. power delivered to the load. It is denoted by η as a percentage

$$\eta = \frac{\text{signal power delivered to load}}{\text{d.c. power supplied to output circuit}} \times 100\%$$

$$= \frac{\frac{1}{2} V_m I_m}{V_{cc} I_C} \times 100\%$$

$$= 50 \frac{V_m I_m}{V_{cc} I_C} \% \qquad (2.64)$$

Since the maximum values that V_m and I_m can have, are $V_{cc}/2$ and I_C (which is equal to $V_{cc}/2R_L$), the maximum possible efficiency for a class A amplifier is 25%. In this circuit the load has been assumed to be a resistor. If

Fig. 2.26. Transformer-coupled class A common-emitter amplifier.

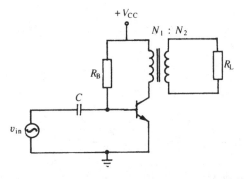

Fig. 2.27. Capacitor-coupled class A common-emitter amplifier.

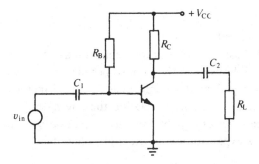

it is not so, and the load has a power factor of $\cos\theta$ then the term $V_m I_m/2$ in both (2.62) and (2.63) should be multiplied by $\cos\theta$.

Other configurations of class A amplifiers are shown in Figs. 2.26 and 2.27. In the circuit of Fig. 2.26 the load is coupled to the circuit via a transformer with a turns-ratio of $N_1:N_2$. The effective load resistance $r_L = (N_1/N_2)^2 R_L$ and the a.c. load line has a slope of $-1/r_L$. In this case there is no dissipation of d.c. power since instead of a resistive load we now have an inductive load, and the maximum possible efficiency is 50%. In the circuit of Fig. 2.27 the load is coupled to the circuit via a capacitor. This time the effective load resistance is the parallel combination of R_C and R_L.

Worked example 2.10
    Given that $V_{CC} = 24$ V and $R_L = 8\,\Omega$, find a suitable transistor for the circuit of Fig. 2.24. What is the maximum possible a.c. power output?

Solution

From (2.63)

$$P_D = V_C I_C - \frac{V_m I_m}{2}$$

P_D is maximum when output power, $\dfrac{V_m I_m}{2} = 0$.

$$\therefore P_{D(max)} = V_C I_C = \frac{V_{CC}}{2} \cdot \frac{V_{CC}}{2R_L} = \frac{24 \times 24}{2 \times 2 \times 8} = 18 \text{ W}$$

∴ The transistor should be able to dissipate 18 W. Maximum a.c. output power is

$$\frac{V_{CC}^2}{8R_L} = \frac{24 \times 24}{8 \times 8} = 9 \text{ W}.$$

2.4.2   Class B amplifiers

In the case of Class A amplifiers we assumed that a transistor is a linear device. In general, however, this is not so because the static output characteristics are not equidistant straight lines for equal increments of input excitation. Therefore, instead of relating the alternating collector current i_{CQ} with the base current i_{BQ} by the linear equation $i_{CQ} = ki_{BQ}$, we may use a power-series expansion and write more accurately

$$i_{CQ} = k_1 i_{BQ} + k_2 i_{BQ}^2 + k_3 i_{BQ}^3 + k_4 i_{BQ}^4 \qquad (2.65)$$

where *k*s are constants.

Now let us assume that the input waveform is sinusoidal and given by

$$i_{BQ} = I_{bm} \cos \omega t \qquad (2.66)$$

then the collector current can be found from (2.65) and (2.66), by using trigonometric transformations, to be

$$i_{CQ} = A_0 + A_1 \cos \omega t + A_2 \cos 2\omega t + A_3 \cos 3\omega t + \dots \qquad (2.67)$$

The total instantaneous current will therefore be

$$i_C = I_C + i_{CQ} = I_C + A_0 + A_1 \cos \omega t + A_2 \cos 2\omega t + A_3 \cos 3\omega t \quad (2.68)$$

The values of the *A* terms can be calculated in terms of *k*s and hence *A*s are also constants.

Therefore, the non-linear characteristic of the transistor introduces into the output components whose frequencies are multiples of that of the

Fig. 2.28. A class B push–pull amplifier.

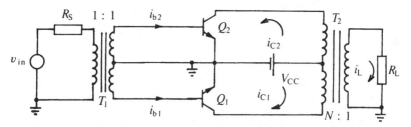

sinusoidal input excitation. It also brings in an additional d.c. component A_0 thus yielding a total d.c. component of collector current $I_C + A_0$.

The class B push–pull amplifier shown in Fig. 2.28 can eliminate a large portion of distortion introduced by the second harmonic components if the stages are perfectly matched. The input signal is applied through a centre-tapped transformer T_1 which supplies two base currents of equal amplitudes but 180° out of phase for transistors Q_1 and Q_2.

Let us assume that the base current to transistor Q_1 is

$$i_{b1} = I_{bm} \cos \omega t$$

Then the base current to transistor Q_2 will be of the form

$$i_{b2} = I_{bm} \cos(\omega t + \pi)$$

Using (2.68), we may write

$$i_{C1} = I_C + A_0 + A_1 \cos \omega t + A_2 \cos 2\omega t + A_3 \cos 3\omega t \qquad (2.69)$$

and

$$i_{C2} = I_C + A_0 - A_1 \cos \omega t + A_2 \cos 2\omega t - A_3 \cos 3\omega t \qquad (2.70)$$

These two collector currents flow in opposite directions through the primary windings of the output transformer T_2. It can be seen from Fig. 2.28 that the load current i_L is directly proportional to the difference of these collector currents and assuming that the two transistors are identical we have

$$i_L = N(i_{C1} - i_{C2}) = 2N(A_1 \cos \omega t + A_3 \cos 3\omega t) \qquad (2.71)$$

Thus a push–pull arrangement balances out all even harmonics in the output. However, if the characteristics of the two transistors are not the same then even harmonics will appear in the output.

The load current of the class B amplifier shown in Fig. 2.28 is sinusoidal, but in practice, is distorted near the zero-crossing, as shown in Fig. 2.29.

Fig. 2.29. Load current of a class B push–pull amplifier showing crossover distortion.

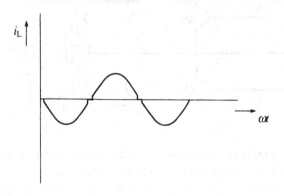

Fig. 2.30. Push–pull amplifier with bias supply: (*a*) battery supply; (*b*) resistance-divider supply.

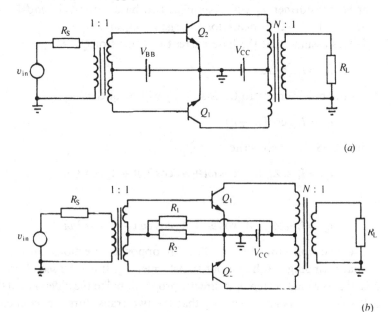

This effect, known as *crossover distortion*, is due to the fact that the linear operation of a transistor begins only when the base current is positive enough for base-emitter voltage v_{BE} to exceed the base-emitter 'on' voltage, which is usually 0.7 V for silicon transistors. The crossover distortion may be eliminated by biasing the two transistors at approximately 0.7 V using the techniques shown in Fig. 2.30.

Fig. 2.31. Output waveforms of a single stage class B amplifier.

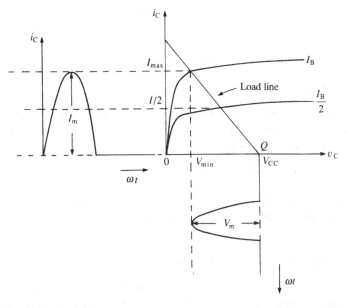

Power calculations

The output power of the circuit can be determined by observing the output voltage and current waveforms of a single stage class B amplifier as shown in Fig. 2.31 and taking into consideration that the amplifier as a whole produces sinusoidal output waveforms. It is given by

$$P = I_m V_m / 2 = I_m (V_{CC} - V_{min}) / 2 \qquad (2.72)$$

The power delivered by the d.c. supply is found by taking the product of the power supply voltage and the d.c. current flowing into the circuit. The current flowing from the supply is the sum of i_{C1} and i_{C2}. From Fig. 2.32, we

Fig. 2.32. Power supply current waveform of a class B push–pull amplifier.

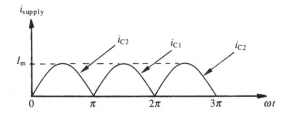

may observe that this is a full-wave rectified current. The d.c. supply current is just the average value of this waveform or

$$I_{d.c.} = 2I_m/\pi \qquad (2.73)$$

Thus the d.c. input power from the supply is

$$P_i = 2I_m V_{CC}/\pi \qquad (2.74)$$

The input power to the circuit is accounted for by the dissipation of both transistors and the output power. From (2.72) and (2.74) we find the transistor dissipation to be

$$P_D = P_i - P = 2I_m V_{CC}/\pi - I_m V_m/2$$

But $I_m = V_m/R'_L$ where $R'_L = N^2 R_L$, and $2N$ is the ratio of the number of primary turns to the number of secondary turns.

Thus

$$P_D = 2V_m V_{CC}/\pi R'_L - V_m^2/2R'_L \qquad (2.75)$$

We may observe from (2.75) that the transistor dissipation is zero in the absence of any input signal, since V_m will then be equal to zero. As the input signal increases the peak output voltage, V_m rises, and hence the transistor dissipation increases. For maximum value of the transistor dissipation we set

$$\frac{dP_D}{dV_m} = \frac{2V_{CC}}{\pi R'_L} - \frac{2V_m}{2R'_L} = 0$$

or

$$V_m = 2V_{CC}/\pi$$

Substituting this value in (2.75) we obtain

$$P_{D(max)} = \frac{2.2V_{CC}^2}{\pi^2 R'_L} - \frac{4V_{CC}^2}{2\pi^2 R'_L} = \frac{2V_{CC}^2}{\pi^2 R'_L} \qquad (2.76)$$

The maximum output power is obtained by assuming $V_m = V_{CC}$ (i.e. $V_{CC} \gg V_{min}$). Therefore,

$$P_{max} = \frac{V_{CC}^2}{2R'_L} \qquad (2.77)$$

So we may write from (2.76) and (2.77)

$$P_{D(max)} = \frac{4}{\pi^2} P_{max} \simeq 0.4 P_{max} \qquad (2.78)$$

Now the collector efficiency of the amplifier, from (2.72) and (2.74), is

$$\eta = \frac{P}{P_i} \times 100\% = \frac{\dfrac{I_m}{2}(V_{CC} - V_{min})}{2\dfrac{I_m}{\pi} V_{CC}} \times 100\%$$

$$= \frac{\pi}{4}\left(1 - \frac{V_{min}}{V_{CC}}\right) \times 100\%. \tag{2.79}$$

If $V_{CC} \gg V_{min}$, then

$$\eta \simeq \frac{\pi}{4} \times 100\% = 78.5\%.$$

Therefore, the maximum attainable efficiency for a class B amplifier is 78.5% compared with 50% for transformer-coupled class A amplifiers.

Worked example 2.11

The peak collector current and voltage in transistors Q_1 and Q_2 of Fig. 2.28 are 3 A and 14 V, respectively. Calculate (a) the output power of the circuit, (b) the average power supplied by the d.c. source, (c) the average power dissipated by each transistor, and (d) the efficiency. Given,

$$V_{CC} = 28 \text{ V}$$

Solution

(a) From (2.72)

$$P = \frac{V_m I_m}{2} = \frac{14 \times 3}{2} = 21 \text{ W}$$

(b) From (2.74)

$$P_i = \frac{2 I_m V_{CC}}{\pi} = \frac{2 \times 3 \times 28}{\pi} = 53.5 \text{ W}$$

(c) Power dissipated by two transistors

$$53.5 - 21 = 32.5 \text{ W}$$

∴ average power dissipated by each transistor is

$$\frac{32.5}{2} = 16.25 \text{ W}.$$

(d)   From (2.79)

$$\eta = \frac{\pi}{4}\left(1 - \frac{V_{min}}{V_{CC}}\right) \times 100\%$$

$$= \frac{\pi}{4}\left(1 - \frac{V_m}{V_{CC}}\right) \times 100\% = \frac{\pi}{4}\left(1 - \frac{14}{28}\right) \times 100\% = 39.25\%$$

2.4.3   Class C amplifiers

Class C amplifiers are, in general, used for amplification of a single frequency or over a very narrow frequency band. The load current which exists for less than one-half of the cycle, generates a sinusoidal output voltage by flowing through a resonant circuit tuned to the fundamental frequency or one of the harmonic components. Figure 2.33 shows the class C amplifier as it is normally operated, with a resonant network in the collector circuit. The resonant frequency of the network is approximated by

$$f_o \simeq 1/2\pi\sqrt{(LC)} \qquad\qquad (2.80)$$

if the resistance of the coil is assumed to be small. The transistor conducts only when the input voltage exceeds the sum of the negative bias and the base-emitter 'on' voltage of the transistor. The amplitude of the fundamental frequency component of the output waveform is determined by the angle θ_C during which the transistor conducts. By observing Fig. 2.34 we may write that the conduction angle,

$$\theta_C = 2\cos^{-1}\frac{V_{BB} + 0.7}{V} \qquad\qquad (2.81)$$

Fig. 2.33. A class C amplifier.

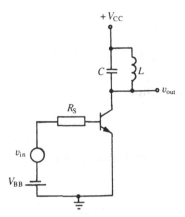

Fig. 2.34. Input voltage and output current of a class C amplifier. θ_C depends on V and V_{BB}.

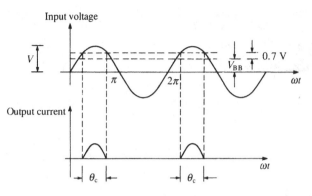

where V_{BB} is the bias voltage and V is the peak input voltage.

The efficiency of this amplifier is large because the transistor is cut-off during most of every full cycle of the input, thus dissipating a very small amount of power. In fact, as the conduction angle approaches zero, the efficiency of the amplifier approaches 100%. But at the same time the output power also tends toward zero. So a compromise between high efficiency and large output power is made and a typical efficiency of 80% can be achieved.

Worked example 2.12

The power supply voltage, bias voltage and the peak value of the input voltage for the circuit of Fig. 2.33, are 28 V, 5 V and 8 V respectively. Find the conduction angle. If $C = 100\,\text{pF}$, find also the inductance necessary to tune the amplifier to the frequency of 2 MHz.

Solution
From (2.81)

$$\theta_C = 2\cos^{-1}\frac{5+0.7}{8} = 89°$$

From (2.80)

$$f_o = \frac{1}{2\pi\sqrt{(LC)}}$$

or

$$L = \frac{1}{(2\pi f_o)^2 C} = \frac{1}{(2\pi \times 2 \times 10^6)^2 \times 100 \times 10^{-12}} = 0.063\,\text{mH}$$

Fig. 2.35. (a) Transistor mounted on a heat sink in free air and (b) the electrical analog of the thermal system.

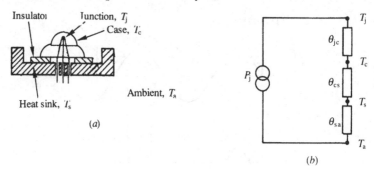

(a)

(b)

2.4.4   Heat sinks

These are used to keep the power transistor junction below some maximum specified operating temperature which is typically 100 °C for germanium transistors and 150–200 °C for silicon transistors. The power transistors are normally bolted to the metal heat sinks, sometimes separated by insulators which provide very high insulation resistance and high thermal conductivity. The heat is conducted outward to metal fins of the heat sinks from which convection and radiation into the air take place, see Fig. 2.35. For a power transistor mounted on a heat sink the junction temperature is dependent on the power being dissipated by the transistor, the thermal conductivity of the transistor case, the style of the heat sink and also the ambient temperature. Usually manufacturers specify the maximum junction temperature of a transistor along with the thermal resistance between the collector junction and the exterior of the case. The thermal resistance, θ, is defined as the ratio of the temperature rise to the power dissipated. An increase in the junction temperature, T_j, above the case temperature, T_c, is related to the power dissipated, by the equation

$$T_j - T_c = \theta_{jc} P_j$$

where

$$T_j - T_c = \text{temperature rise in } °C$$

$$P_j = \text{power dissipated at junction in watts}$$

and

$$\theta_{jc} = \text{thermal resistance between junction and case in } °C/\text{watt}$$

Similarly the case-to-heat sink (including insulator) thermal resistance in °C/watt,

$$\theta_{cs} = (T_c - T_s)/P_j$$

Fig. 2.36. Typical examples of heat sinks. (RS Components Ltd.)

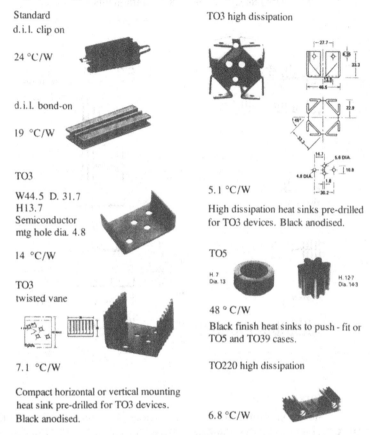

Standard
d.i.l. clip on

24 °C/W

d.i.l. bond-on

19 °C/W

TO3

W44.5 D. 31.7
H13.7
Semiconductor
mtg hole dia. 4.8

14 °C/W

TO3
twisted vane

7.1 °C/W

Compact horizontal or vertical mounting
heat sink pre-drilled for TO3 devices.
Black anodised.

TO3 high dissipation

5.1 °C/W

High dissipation heat sinks pre-drilled
for TO3 devices. Black anodised.

TO5

48 ° C/W

Black finish heat sinks to push - fit or
TO5 and TO39 cases.

TO220 high dissipation

6.8 °C/W

and the heat sink-to-ambient thermal resistance in °C/watt,

$$\theta_{sa} = (T_s - T_a)/P_j$$

The circuit shown in Fig. 2.35 represents an electrical analog of the thermal
system of a transistor mounted on a heat sink in free air. By examining the
figure we may write

$$T_j = P_j(\theta_{jc} + \theta_{cs} + \theta_{sa}) + T_a \qquad (2.82)$$

where T_a is the ambient temperature in °C.

All power transistors are packaged in cases which permit good physical
contact with a heat sink. In most power transistors the case is connected to
the collector. The case-to-heat sink thermal resistance depends on the type
of package. The sink-to-ambient thermal resistance depends mainly on the
surface area of the heat sink. Figure 2.36 shows typical examples of heat
sinks and Table 2.2 gives the values of the thermal resistances.

Table 2.2. *High power finned heat sink.* (RS Components Ltd.)

High power range			
Finned heat sinks, black finished (including ends). Thermal resistance quoted is with fins vertical in free air.			
thermal resistance	section	thermal resistance	section
0.23 °C/W L. 250 W. 235 H. 100		1.1 °C/W L. 152 W. 130 H. 32	 Supplied with four 4 B.A. nuts to aid mounting
0.3 °C/W L. 250 W. 300 H. 40		1.3 °C/W L. 87.5 W. 108 H. 58.7	
0.41 °C/W L. 250 W. 200 H. 40	 Will accept RS surface mounted solid state relays.	1.7 °C/W L. 100 W. 98.4 H. 53.2	
0.5 °C/W L. 115 W. 120 H. 120		2.1 °C/W L. 124 W. 124 H. 26.7	
0.65 °C/W L. 150 W. 163.5 H. 58.7		2.5 °C/W L. 75 W. 104.8 H. 25.9	
0.75 °C/W L. 100 W. 114.3 H. 114.3		3.0 °C/W and 3.5 °C/W	 L. 87.5 (3 °C/W) W. 108 H. 14 L. 75 (3.5°C/W) W. 108 H. 14
0.9 °C/W L. 125 W. 200 H. 25		4.0 °C/W L. 100 W. 64.5 H. 15	

Worked example 2.13

The maximum collector dissipation for a push–pull amplifier is 32 W. What would be the junction temperature if the ambient temperature is 38 °C and the thermal resistance of the heat sink used is 3.2 °C/W? Given, $\theta_{jc} = 1.8$ °C/W and $\theta_{cs} = 0.3$ °C/W.

Solution

Since a push–pull amplifier has two transistors, each of them will have maximum collector dissipation of $\frac{32}{2} = 16$ W. Now from (2.82)

$$T_j = P_j(\theta_{jc} + \theta_{cs} + \theta_{sa}) + T_a$$

$$= 16(1.8 + 0.3 + 3.2) + 38$$

$$= 122.8 \,°C$$

Worked example 2.14

The maximum allowable junction temperature of the power transistor 2N3055 is 200 °C and its junction-to-case thermal resistance is

1.5 °C/W. It is to be used in an ambient temperature of 30 °C. If the maximum power dissipation of the transistor is 25 W, and the insulator used between the heat sink and the transistor has a thermal resistance of 0.2 °C/W, what type of heat sink should be used?

Solution

Choose $T_j = \frac{2}{3} \times$ maximum allowable temperature. From (2.82) we then have

$$T_j = P_j(\theta_{jc} + \theta_{cs} + \theta_{sa}) + T_a$$

or,

$$\tfrac{2}{3} \times 200 = 25(1.5 + 0.2 + \theta_{sa}) + 30$$

$$\therefore \; \theta_{sa} = \frac{\dfrac{2 \times 200}{3} - 30}{25} - 1.7 = 2.43\ °\text{C/W}$$

Choose a heat sink (RS stock no 403-140) with $\theta_{sa} = 2.5\ °\text{C/W}$.

Summary

1. For a differential amplifier the common-mode rejection ratio gives a qualitative measure of the ability of a circuit to respond to difference signals while being insensitive to common-mode signals, such as noise. It is the ratio of the differential-mode voltage gain to the common-mode voltage gain. The differential-mode voltage gain is the ratio of the output voltage to the difference between the two input voltages, whereas, the common-mode voltage is the ratio of the output voltage to the voltage applied simultaneously to both input terminals.

2. In order to source current in integrated circuits current mirrors are widely used. In integrated circuit operational amplifiers, as an example, the operating current in the whole circuit can be set by a single resistor.

3. The Darlington connection of two transistors acts as a single transistor with a very high gain and finds use mainly in power electronics. It is also a high input impedance device and is used as an emitter follower.

4. The importance and design of level shifting circuits especially for IC have been discussed.

5. An analysis of an amplifier with two stages in cascade has been made in Section 2.2. Ways of choosing the right transistor amplifier configuration for a particular stage have been discussed. The lower and upper cutoff frequencies of a cascaded amplifier with n identical stages are given by $f_L/\sqrt{(2^{1/n}-1)}$ and $f_U[\sqrt{(2^{1/n}-1)}]$ respectively, where f_L and F_U are the lower and upper cutoff frequencies of a single stage amplifier.

6. The maximum possible efficiency of a transformer-coupled class A power amplifier is 50%, whereas, that of a class B push–pull power amplifier is 78.5%. Another advantage of the class B amplifier is that it eliminates distortion due to even harmonics.

7. Power transistors are usually mounted on heat sinks in order to avoid overheating of their junctions. The overheating shortens the life of a transistor. The right type of heat sink can be chosen using the expression, $T_j = P_j(\theta_{jc} + \theta_{cs} + \theta_{sa}) + T_a$ where the parameters have the usual meanings.

Problems

2.1. What is the collector current of the ideal differential amplifier shown in Fig. 2.1 if $R_c = 2.2 \, \text{k}\Omega$, $R_E = 1.5 \, \text{k}\Omega$ and the common-mode rejection ratio is 60? If the input terminal B_1 is grounded and a sinusoidal signal is applied to the input terminal B_2, then which output terminal would give a non-inverting gain? If the output voltage of the amplifier at that terminal is 1.6 V peak-to-peak, what is the value of the input signal? Assume that the circuit works in an environment having a temperature of $27 \, °C$.

2.2. Transistors Q_1 and Q_2 of the circuit shown in Fig. ex.2.2 are matched and each has β equal to 90. The single ended differential voltage gain is 80 and the common-mode rejection ratio is 6400. Find (a) the double ended differential gain, (b) the collector output resistance of Q_3 and (c) the emitter current of Q_3. Assume that the circuit works at a temperature of $20 \, °C$.

2.3. All the transistors in the circuit shown in Fig. ex.2.3 are identical, each having a current gain of 100. Determine the collector current of the transistors if $R = 4.7 \, \text{k}\Omega$. The base-emitter 'on' voltage of the transistors is 0.7 volts.

2.4. Transistor Q_3 of the circuit shown in Fig. P.2.4 has $\beta_3 = 100$. Assuming that Q_1 and Q_2 are matched, find approximate values for (a) the emitter currents in Q_1 and Q_2, and (b) the d.c. output voltages V_{o1} and V_{o2}. Assume $V_{BE(ON)} = 0.7 \, \text{V}$.

2.5. Transistors Q_1, Q_2, Q_3 and Q_4 in the circuit of Fig. ex.2.4 are all identical. Find the current gain of the transistors if the d.c. output voltage at the terminals B_1 and B_2 is 6.8 volts. Assume $V_{BE(ON)} = 0.7 \, \text{V}$.

2.6. Assuming perfectly matched transistors, find the approximate values of I_C and V_{CE} in each of transistors Q_1, Q_2 and Q_3 in Fig. P.2.6.

2.7. When three transistors are biased from a single current source, such as shown in Fig. P.2.6, show that the ratio of the collector current I_C to the

Fig. P.2.4

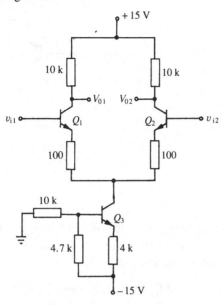

Fig. P.2.6

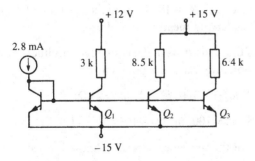

source current I, is given by

$$I_C/I = 1/(1 + 4/\beta)$$

where β is the current gain. Assume that all transistors are perfectly matched.

2.8. In the current mirror circuit shown in Fig. P.2.8, the transistors have a V_{BE} offset of 3 mV. If the βs of both devices are 20, calculate the current mirror ratio I_2/I_1 at a temperature of 17 °C.

Fig. P.2.8   Fig. P.2.10

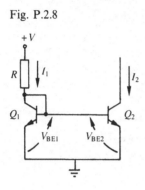
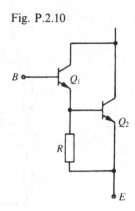

2.9. In a Darlington pair, the current gains of the two transistors are 60 and 80. Find the collector current if the effective input resistance of the transistor is 200 kΩ at a room temperature of 30 °C.

2.10. The gains of transistors Q_1 and Q_2 in a Darlington power transistor shown in Fig. P.2.10 are 120 and 80 respectively. Find the quiescent collector current when the Darlington pair is on. Assume that the resistor has a value of 100 Ω and the base-emitter 'on' voltage for Q_2 is 0.7 volt.

2.11. In the circuit of Fig. ex.2.7, R_1 is 18 kΩ. Find the d.c. level of the input signal for output voltages having mean values of zero.

2.12. The parameters of the transistors in the two-stage amplifier shown in Fig. 2.16, at the corresponding quiescent points are:

$$h_{oc} = 3 \times 10^{-5} \text{ S}; \ h_{fc} = -100; \ h_{rc} = 1; \ h_{ic} = 2 \text{ k}\Omega;$$

$$h_{oe} = 8 \times 10^{-5} \text{ S}; \ h_{fe} = 100; \ h_{re} = 3 \times 10^{-4}; \ h_{ie} = 1.5 \text{ k}\Omega$$

If the output resistance of the amplifier is 60 Ω, what is the value of R_S? Given $R_C = R_E = 5.6 \text{ k}\Omega$.

2.13. In the circuit of Fig. 2.19, the effective Q is 50 and the bandwidth of the amplifier is 20 kHz. If the output resistance of the transistor is 50 kΩ, $L = 80 \ \mu\text{H}$ and $\beta = 85$, find (a) the resonant frequency and (b) the Q-factor.

2.14. Find a suitable transistor for the circuit of Fig. 2.24, if $V_{CC} = 12$ volts and $R_L = 3 \ \Omega$. What is the maximum possible a.c. power output?

2.15. The class-A amplifier in Fig. P.2.15 is biased at $I_C = 0.2$ A. The transformer resistance is negligible. (a) What is the slope of the a.c. load line? (b) At what value does the a.c. load line intersect the V_{CE}-axis?

Fig. P.2.15   Fig. P.2.17

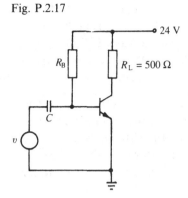

(c) What is the maximum peak-value of the collector voltage without distortion? (d) What is the maximum power delivered to the load under the conditions of (c)? (e) What is the amplifier efficiency under the conditions of (c)?

2.16. The average power dissipated by each of transistors Q_1 and Q_2 of Fig. 2.28 is 16 watts and the efficiency of the amplifier is 40%. Calculate (a) the peak collector voltage and (b) the peak collector current if $V_{CC} = 24$ volts.

2.17. The amplifier shown in Fig. P.2.17 is biased at $I_Q = 20$ mA. Find (a) the a.c. power in the load resistance when the voltage swing is the maximum possible without distortion and (b) the amplifier efficiency under the conditions of (a).

2.18. The peak value of the input voltage and the conduction angle of a class C amplifier are 6 volts and 90° respectively. Find the bias voltage, if $L = 60\,\mu$H. Find also the capacitance necessary to tune the amplifier to the frequency of 1 MHz.

2.19. A push–pull amplifier has a junction temperature of 70 °C, what would be the maximum collector dissipation for the amplifier if the thermal resistance of the heat sink used is 2.6 °C/W and the ambient temperature is 30 °C? Given, $\theta_{jc} = 2$ °C/W and $\theta_{cs} = 0.4$ °C/W.

2.20. The thermal resistance between a semiconductor device and its case is 0.8 °C/W. It is used with a heat sink whose thermal resistance to ambient is 0.5 °C/W. If the device dissipates 10 W and the ambient temperature is 47 °C, what is the maximum permissable thermal resistance between the case and the heat sink? The device temperature can not exceed 70 °C.

2.21. A semiconductor device dissipates 25 W through its case and its heat sink to the surrounding air. The thermal resistances are the following: device-to-case, 1 °C/W; case-to-heat sink, 1.2 °C/W; and heat sink-to-air, 0.7 °C/W. In what maximum air temperature can the device be operated if its temperature can not exceed 100 °C?

2.22. What is the junction temperature of a power transistor if its junction-to-case thermal resistance is 3.0 °C/W, the maximum power dissipation of the transistor is 20 W, the insulator used between the heat sink and the transistor has a thermal resistance of 0.8 °C/W, the ambient temperature is 30 °C and the heat sink-to-air thermal resistance is 2.2 °C/W?

3

Operational amplifiers

Objectives
At the end of the study of this chapter the student should be:

1. familiar with the main imperfections in operational amplifiers, able to describe their effects on the output and stability of various circuits, and compensate for the errors due to them.
2. able to design several important linear and nonlinear circuits using operational amplifiers: phase shifting circuits, instrumentation amplifiers, comparators, precision rectifiers and logarithmic amplifiers.
3. familiar with different methods of design of active filters and able to choose the right design for a particular need.
4. able to describe the operating principles of multivibrators and triangular wave generators, and design these circuits given the specifications.
5. able to solve differential equations using summers, integrators and potentiometers, and apply amplitude and time scaling if necessary.
6. familiar with the principles of inverse function generators and able to design dividing, square rooting and RMS circuits using multipliers and operational amplifiers.

The name operational amplifier is derived from the fact that the amplifier was originally used to perform electronically various mathematical operations such as differentiation, integration, addition and subtraction. However, due to its versatility its use has been extended to other types of electronic circuits mainly in the fields of instrumentation and control engineering. The availability of inexpensive high performance operational amplifiers in the form of integrated circuits has obviously extended their use especially in analogue electronic circuits and systems. In this chapter it will be most appropriate for us to consider first the main imperfections in operational amplifiers which affect the performance of a circuit and how to deal with them. Later in the chapter various applications will be described in some detail.

3.1   Imperfections in operational amplifiers

Ideally an operational amplifier should have infinite gain, infinite input impedance, zero output impedance and infinite bandwidth, but in practice it does not have these characteristics due to the conservation of energy in its internal circuitry. The main imperfections in an operational amplifier which stop a circuit behaving ideally and may produce significant errors and instability if not properly compensated are (a) slew rate, (b) input offset voltage, (c) bias and offset currents and (d) frequency response effects. We will consider these imperfections one by one and when the effect of one is considered the effects of others will be ignored for simplicity.

3.1.1   Slew rate and its effect on full power bandwidth

An operational amplifier does not respond instantly to a sudden change in the input voltage due to its internal capacitances (frequency compensating or stray capacitances).* These can only be charged up at a constant rate which is again limited by the current available, thus slowing down the speed of response of the circuit. The slew rate is defined as the maximum rate of change of output voltage an operational amplifier can give per unit time and it is usually expressed in volts/microsecond.

$$S = \text{slew rate} = (\Delta v_0/\Delta t)_{max}$$

Practical integrated circuit operational amplifiers have specified slew rate ranging from $0.1\,\text{V}/\mu\text{s}$ to $500\,\text{V}/\mu\text{s}$.

The slew rate is also a measure of the ability of an operational amplifier to amplify a sinusoidal input signal without distortion. If an operational amplifier is connected to power supplies such that its maximum output voltage swing is $\pm V_{op}$ then for a sinusoidal signal of frequency, f, the instantaneous output voltage, v, is given by

$$v = V_{op} \sin 2\pi ft \tag{3.1}$$

In order to obtain the slope of the output voltage we differentiate the instantaneous value v

$$dv/dt = 2\pi f V_{op} \cos 2\pi ft \tag{3.2}$$

Now to find the maximum slope, we differentiate again and equate the result to zero

$$-4\pi^2 f^2 V_{op} \sin 2\pi ft = 0$$

---

* Frequency compensating capacitors are mainly used in order to obtain stable circuits at relatively low gain, it will be discussed later in this section.

Fig. 3.1. Effect of slew rate on the output waveform for (a) a step input signal and (b) a sinusoidal input signal having a frequency much higher than $S/2\pi V_{op}$.

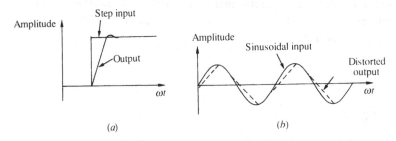

or,

$$2\pi ft = 0$$

(since f and V_{op} have finite values). Substituting this in (3.2) we obtain

$$(\mathrm{d}v/\mathrm{d}t)_{max} = 2\pi f V_{op} \tag{3.3}$$

This maximum slope should be equal to or less than the slew rate of the operational amplifier to avoid any distortion occurring in the output waveform. Therefore we may write

$$f \leqslant S/2\pi V_{op} \tag{3.4}$$

where f is the full power bandwidth of the circuit.

Figure 3.1 shows the effect of slew rate on the output waveforms for a step input signal and also a sinusoidal input signal having a frequency much higher than $S/2\pi V_{op}$. High speed operational amplifiers (such as type RS5539 with a slew rate of 600 V/μs and unity gain-bandwidth of 1.2 GHz) are used where high rates of output voltage change are required.

Worked example 3.1

An inverting amplifier has a gain of 100 and the operational amplifier used in the circuit has a slew rate of 1.5 V/μs. If a step input of 100 mV is fed to the circuit, calculate the time required for the output to reach within 1% of its final value.

Solution

The output voltage $= -100 \times 100\,\mathrm{mV} = -10\,\mathrm{V}$

Since the output of the operational amplifier changes by an amount 1.5 V in 1 μsec, a change of $(10 - 0.1) = 9.9$ V occurs in 6.6 μsec.

Worked example 3.2

The slew rate of the 741 type operational amplifier is typically 0.5 V/μs. If it is used in the circuit mentioned in Ex. 3.1 and the input is a sinusoidal signal of 35 kHz, what is the maximum amplitude the output can have without any distortion?

Solution

From (3.4)

$$f \leqslant S/2\pi V_{op}$$

Substituting for $f = 35$ kHz, $S = 0.5$ V/μsec and solving for V_{op}

$$V_{op} = \frac{S}{2\pi f} = \frac{0.5 \times 10^6}{2\pi \times 35 \times 10^3}$$

$$= 2.275 \text{ V}$$

Therefore the maximum amplitude of the input signal

$$V_{ip} = \frac{2.275}{100} = 22.75 \text{ mV}$$

3.1.2 Input offset voltage

In ideal operational amplifiers the output should be zero in the absence of any input signal across the differential inputs as illustrated in Fig. 3.2(*a*). However, in practice due to various imbalances (e.g., mismatches in the V_{BE}s of the transistors in the differential stages) in the internal circuitry of the operational amplifier, there is always a voltage present at the output even when both the input terminals are connected to ground. This output voltage can be set to zero by applying a voltage across the input terminals. This voltage known as the input offset voltage and denoted by V_{io} may be either positive or negative. The typical value of V_{io} for the most commonly used operational amplifier type 741 is 2 mV.

A practical operational amplifier can be modelled as shown in Fig. 3.2(*b*). The d.c. source representing the input offset voltage and connected in series with either of the two input terminals with the correct polarity can be used to find the unwanted output due to V_{io}. If we now consider an inverting amplifier with an input resistor R_1 and a feedback resistor R_2 and connect terminal 'A' to ground as shown in Fig. 3.2(*c*), then it is apparent from the diagram that the output offset voltage due to V_{io} at the noninverting input is given by

$$V_{os} = V_{io} \cdot (R_1 + R_2)/R_1 \tag{3.5}$$

Fig. 3.2(*a*). Offset input voltage of a practical operational amplifier.

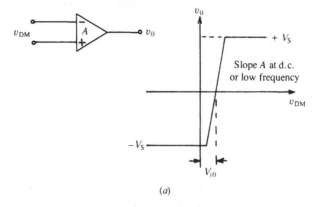

(*a*)

Fig. 3.2(*b*). A model of a practical operational amplifier taking into account of the offset input voltage.

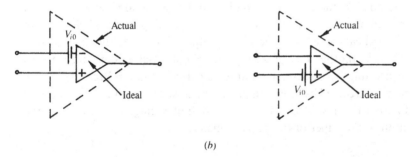

(*b*)

Fig. 3.2(*c*). An inverting amplifier with a d.c. source representing the input offset voltage in series with the noninverting terminal.

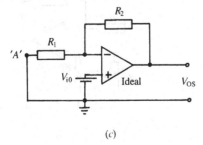

(*c*)

Fig. 3.2(d). An inverting amplifier with a d.c. source representing the input offset voltage in series with the inverting terminal.

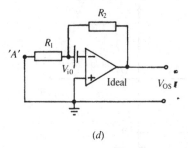

(d)

Therefore the circuit amplifies the input offset voltage by a factor equal to the closed loop gain of the amplifier connected in the noninverting configuration. It can be shown that (3.5) is also valid for an amplifier circuit when the d.c. source representing the input offset voltage V_{io} is in series with the inverting terminal (Fig. 3.2(d)).

We can compensate for the input offset voltage by using several techniques. One of these is to connect a potentiometer between the offset null terminals of the integrated circuit chip. The potentiometer wiper is then connected to a d.c. supply voltage of required polarity. This type of adjustment can be made in the 741 type operational amplifier as shown in Fig. 3.3(a). Another way of setting the output offset voltage to zero is to apply a voltage to one of the input terminals as shown in Fig. 3.3(b). This voltage is magnified by a factor equal to the closed loop gain of the amplifier and added to or subtracted from the output voltage depending on the position of the wiper of the potentiometer.

Fig. 3.3(a). Compensation of input offset voltage by applying a voltage between the null terminals of the IC chip.
(b). Compensation of input offset voltage by applying a voltage to one of the input terminals.

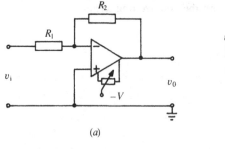

(a)

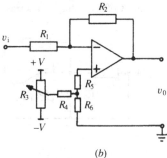

(b)

Worked example 3.3

The maximum input offset voltage for the operational amplifier in the circuit of Fig. 3.2(c) is 7.5 mV. If $R_1 = 1$ kΩ and $R_2 = 47$ kΩ, what will be the worst case offset voltage at the output?

Solution

From (3.5) we have

$$V_{os} = V_{io}\left(1 + \frac{R_2}{R_1}\right)$$

we substitute 7.5 mV for V_{io} and find

$$V_{os} = 7.5 \times 10^{-3}\left(\frac{1 \times 10^3 + 47 \times 10^3}{1 \times 10^3}\right) = 0.36 \text{ V}$$

3.1.3 Bias and offset currents

The input impedance of ideal operational amplifiers should be infinite, thus disabling any current from flowing into the input terminals. However, as we saw earlier, differential amplifiers form the input stage of operational amplifiers, and base currents to the transistors are needed to flow for proper circuit operation. The bias currents flowing into the noninverting and inverting terminals are denoted by I_b^+ and I_b^- respectively. Two current generators connected from the inputs of an ideal operational amplifier to ground can model bias currents I_b^+ and I_b^- as shown in Fig. 3.4(a). These currents may produce an unwanted voltage at the output of an amplifier circuit.

Fig. 3.4(a). A model of a practical operational amplifier taking into account the bias currents.

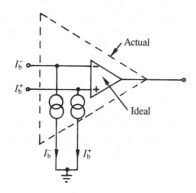

(a)

Fig. 3.4(*b*). Introduction of the compensating resistor to reduce the effect of bias currents on the output.

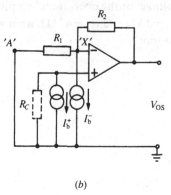

(*b*)

Let us consider an inverting amplifier as shown in Fig. 3.4(*b*). In order to find the effect of bias currents at the output we connect terminal '*A*' to ground. Using Kirchhoff's current law at node *X*, we may write

$$-\frac{V_x}{R_1} + \frac{V_{os}-V_x}{R_2} - I_b^- = 0$$

or,

$$V_{os} = I_b^- R_2 + V_x \frac{R_1+R_2}{R_1} \tag{3.6}$$

or,

$$V_{os} = I_b^- R_2$$

since the differential mode voltage, V_x in this case, is negligible for an ideal operational amplifier. Now if we introduce a compensating resistor R_C between the noninverting terminal and ground, the voltage across the resistor will be equal to $I_b^+ \cdot R_C$ and therefore V_x will be equal to $I_b^+ \cdot R_C$. Substituting this value in (3.6) we obtain

$$V_{os} = I_b^- R_2 + I_b^+ R_C \left(\frac{R_1+R_2}{R_1} \right) \tag{3.7}$$

Since I_b^+ is approximately equal to I_b^- within 20%, (typical values of I_b are 0.1 μA and 0.1 nA for BJT and FET long-tail-pair respectively), in practice, V_{os} can be made negligible by making $R_C = R_1 R_2/(R_1 + R_2)$.*

---

* $I_b = \dfrac{|I_b^+| + |I_b^-|}{2}$, $I_{os} = |I_b^+| - |I_b^-|$; manufacturers specify I_b and I_{os}.

Fig. 3.4(c). A model showing the input offset current and the compensating resistor.

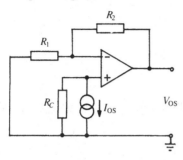

The difference between the two bias currents is called the input offset current and is denoted by I_{os}. It can be modelled as shown in Fig. 3.4(c). Thus the output offset voltage due to the offset current is given by

$$V_{os} = I_{os} \cdot R_2$$

Typically the offset current is about one fifth of the bias current for an operational amplifier. Therefore by introducing a compensating resistor the output offset voltage due to bias currents can be reduced by a factor of five.

Worked example 3.4

The operational amplifier in the circuit of Fig. 3.3(a) has the following specifications: $I_b = 80\,\text{nA}$, $I_{os} = 20\,\text{nA}$ and $V_{io} = 1\,\text{mV}$. If $R_1 = 1\,\text{k}\Omega$ and $R_2 = 47\,\text{k}\Omega$ then calculate the worst case value of the output offset voltage (a) without bias current compensation and (b) with bias current compensation. In case (b) what would be the value of the compensating resistor?

Solution
(a) The output offset voltage due to V_{io} from (3.5)

$$\pm V_{io}(1 + R_2/R_1) = \pm 1 \times 10^{-3}\left(1 + \frac{47 \times 10^3}{1 \times 10^3}\right)$$

$$= \pm 48\,\text{mV}$$

Again $I_b{}^- = (2I_b - I_{os})/2 = (2 \times 80 \times 10^{-9} - 20 \times 10^{-9})/2 = 70\,\text{nA}$. Therefore the output offset voltage due to $I_b{}^-$

$$\pm I_b{}^- R_2 = \pm 70 \times 10^{-9} \times 47 \times 10^3$$

$$= \pm 3.29\,\text{mV}$$

Thus the worst case output offset voltage $= \pm 48\,\text{mV} \pm 3.29\,\text{mV}$

$$= \pm 51.29\,\text{mV}$$

(b) The output offset voltage due to I_{os}

$$\pm I_{os}R_2 = \pm 20 \times 10^{-9} \times 47 \times 10^3$$

$$= \pm 0.94\,\text{mV}$$

Therefore the worst case output offset voltage $= \pm 48\,\text{mV} \pm 0.94\,\text{mV}$

$$= \pm 48.94\,\text{mV}$$

The compensating resistance

$$R_c = \frac{R_1 R_2}{R_1 + R_2} = \frac{1 \times 10^3 \times 47 \times 10^3}{1 \times 10^3 + 47 \times 10^3}$$

$$= 979\,\Omega$$

The output offset voltage due to the input offset voltage and current of an operational amplifier can be compensated almost completely using the techniques described above. However, the input offset voltage and current change with temperature and therefore the maximum compensation can be achieved only at one temperature.

The V_{BE} and β of transistors decrease with increase in temperature and the rates of decrease are not the same for all transistors. These give rise to drifts in offset voltage and offset current respectively.

The input offset voltage and current–temperature coefficients $\Delta V_{io}/\Delta T$ and $\Delta I_{os}/\Delta T$ are given in the manufacturers' specifications of operational amplifiers. For a compensated inverting amplifier the error voltage due to a temperature change can be given by

$$E = \Delta V_{out}/\Delta T(\Delta T) = \left(\frac{\Delta V_{io}}{\Delta T}\right) \cdot \Delta T \cdot (R_1 + R_2)/R_1 + \left(\frac{\Delta I_{os}}{\Delta T}\right) \cdot \Delta T \cdot R_2$$

$$(3.8)$$

Usually the temperature coefficients vary with temperature and they are specified at a particular temperature.

Worked example 3.5

Calculate the error voltage at the output of the circuit of Fig. 3.3(a) due to a temperature change of 30 °C from room temperature. The input offset voltage and current temperature coefficients for the operational amplifier are 3.5 μV/°C and 1.2 nA/°C respectively at room temperature. Given that $R_1 = 2.7\,\text{k}\Omega$ and $R_2 = 100\,\text{k}\Omega$.

Solution

From (3.8) we obtain

$$E = 3.5\,\mu V/°C \times 30\,°C \times \frac{2.7\,k\Omega + 100\,k\Omega}{2.7\,k\Omega}$$

$$+ 1.2\,nA/°C \times 30\,°C \times 100\,k\Omega$$

$$= \left(\frac{3.5 \times 102.7 \times 30}{2.7} + 1.2 \times 30 \times 100\right) \times 10^{-6} = 7.59\,mV$$

Commutating auto-zero operational amplifier

The commutating auto-zero (CAZ) operational amplifier has exceptionally low input offset voltage and low long term input offset voltage drift. It contains basically an oscillator, a counter, analogue switches and two operational amplifiers. In addition to the regular two input terminals it has an auto-zero input terminal. While one of the operational amplifiers processes the input signal, the other is placed in an auto-zero mode and charges a capacitor to a voltage equal to the input offset voltage. The analogue switches reverse the roles of the internal operational amplifiers at a rate designated as the commutation frequency and input offset voltages due to drift with temperature or supply voltages are cancelled out. Figure 3.5 shows a simple noninverting CAZ operational amplifier circuit and the input/output voltage waveform. The RS 76000 CAZ operational amplifier has an input offset voltage of $\pm 5\,\mu V$ maximum and a temperature coefficient of $0.1\,\mu V/°C$. The long term input offset voltage stability of this amplifier is typically $0.2\,\mu V/year$.

Fig. 3.5. A simple noninverting CAZ operational amplifier circuit and its input/output voltage waveforms.

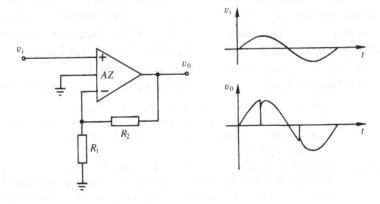

3.1.4　Frequency response effects

The frequency response of an operational amplifier has an important effect on the stability of practical operational amplifier circuits. The stability of a circuit can be determined from the plots of its closed loop gain against frequency and also the open loop gain of the operational amplifier against frequency. Gain versus frequency plots are known as Bode diagrams.

The voltage gain of an ideal operational amplifier should be independent of frequency, but in reality the gain decreases as the frequency increases. This is caused by the stray and semiconductor-junction capacitances present in its circuitry. Operational amplifiers are made up of two or more amplifying stages in cascade and in each stage the distributed capacitances can be assumed to form a single capacitance which in conjunction with the resistances in the circuit becomes responsible for the drop in voltage gain with the increment in frequency.

Let us consider an operational amplifier which has three stages with upper cutoff frequencies at say 200 kHz, 2 MHz and 20 MHz. The gain of each stage reduces at a rate of -20 dB/decade aftet the cutoff frequency.* The magnitude and phase response of the open loop gain for such an operational amplifier is shown in Fig. 3.6. The magnitude response is constant from low frequency (including d.c.) to 200 kHz, thus the bandwidth is 200 kHz at 90 dB. At higher frequencies from 200 kHz to 2 MHz, the gain decreases at a rate of -20 dB/decade. From 2 MHz to 20 MHz the rate of roll off is -40 dB/decade and above 20 MHz it is -60 dB/decade. The unity-gain bandwidth of the operational amplifier circuit is 63.1 MHz at which frequency the zero dB (gain) line intersects the open loop gain curve of the operational amplifier.

In each stage the output voltage lags the input voltage by 90° and for three stages the total phase lag is 270°. In Fig. 3.6 the phase response of the operational amplifier shows that the phase difference between the input and output voltages changes from 0 to $-90°$ for the corner frequency at 200 kHz, from $-90°$ to $-180°$ for the corner frequency at 2 MHz and from $-180°$ to $-270°$ for the corner frequency at 20 MHz.

Now if we connect up this particular amplifier to give an ideal† closed

---

   *   Upper cutoff frequencies are known as corner frequencies in multistage amplifiers. The gains of all stages are multiplied to give the overall gain. Every corner frequency gives an additional rate of roll off of -20 dB/decade.

   †   Loop gain, $A_{OL}\beta(dB) = A_{OL}(dB) - 1/\beta(dB)$, thus the loop gain is the difference in dB of the open loop gain and ideal closed loop gain; in the example the feedback network is resistive, so $1/\beta$ is independent of frequency and $1/\beta$ dB is a straight line parallel to the frequency axis.

Fig. 3.6. The magnitude and phase response of the open loop gain for an operational amplifier.

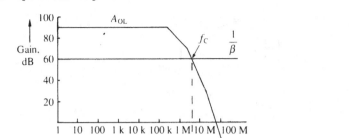

loop gain, $1/\beta$, of 60 dB and superimpose the ideal closed loop gain curve on the open loop gain curve of the operational amplifier then the two curves meet each other at f_c, the crossing frequency. At this frequency the magnitude of the loop gain is zero in dB (i.e., unity)‡ and the phase of the loop gain is 180°. Thus the conditions for oscillations are met and the amplifier circuit is unstable.

By increasing the closed loop gain of the circuit to 70 dB, the phase shift at unity gain can be reduced from $-180°$ to $-135°$ thus giving a phase margin of 45° before the circuit breaks out into sustained oscillations. It is the minimum amount of phase margin (positive) we should allow to keep the circuit free from oscillations.* So we can say that if the ideal closed loop gain, $1/\beta$, curve meets the open loop gain, A_{OL}, curve at a rate of closure of 40 dB/decade on one side and 20 dB/decade on the other side of the intersection, then the circuit is critically stable. Any closed loop gain above this value will give a more stable amplifier circuit but at the expense of bandwidth.

To use this particular operational amplifier at closed loop gains lower than 70 dB, some phase compensation is necessary. Several amplifiers, such

‡ A circuit with negative feedback needs an excess phase shift of 180° only in the feedback loop to turn itself into a positive feedback circuit. Therefore oscillations will occur in such circuits if the phase shift in the loop gain reaches 180° at a frequency at which the magnitude of the loop gain is greater than unity.

* Phase margin = 180° − phase shift at unity gain.

Fig. 3.7. Frequency compensation point for an operational amplifier.

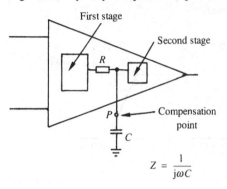

$$Z = \frac{1}{j\omega C}$$

as the type 741, are phase compensated by the manufacturers. This causes its open loop gain curve to roll off at a steady $-20\,\mathrm{dB/decade}$ rate from the first corner frequency to the frequency at which the gain is unity. Therefore the circuit is stable for any closed loop gain since the rate of closure between the closed loop gain curve and the open loop gain curve of the operational amplifier is always $20\,\mathrm{dB/decade}$ on both sides of the intersection and consequently the phase margin is $90°$. Uncompensated amplifiers are provided with terminals to which external components may be connected to stabilise a circuit at relatively low gains.

Lag compensation

Manufacturers usually provide access to an internal point of the operational amplifier for frequency compensation as shown in Fig. 3.7. The resistor R represents the output resistance of the amplifying stage chosen for compensation. When a capacitor, C is connected to the point, P, the RC network introduces a pole in the transfer function of the circuit and thus provides an additional corner frequency. By examining the circuit we can calculate the new open loop gain of the operational amplifier,

$$A_{\mathrm{OL}} = A_{\mathrm{OL}}(\mathrm{old}) \times Z/(R+Z)$$

or,

$$A_{\mathrm{OL}} = A_{\mathrm{OL}}(\mathrm{old}) \times 1/(1 + \mathrm{j}2\pi f C R)$$

or,

$$20\log A_{\mathrm{OL}}\,\mathrm{dB} = 20\log A_{\mathrm{OL}}(\mathrm{old})\,\mathrm{dB} - 20\log(1 + \mathrm{j}2\pi f C R)\,\mathrm{dB} \quad (3.9)$$

The second term on the right hand side of (3.9) thus introduces a pole at $f_x = 1/(2\pi C R)$. Figure 3.8 shows how the compensating network affects the open loop gain curve of an operational amplifier. The uncompensated operational amplifier's open loop gain curve has corner frequencies at f_1 and f_2. From d.c. to the frequency f_1 the gain is constant, after that it drops off at a rate of $-20\,\text{dB/decade}$ up to the second corner frequency f_2 after which the rate of roll off becomes $-40\,\text{dB/decade}$. Now after the introduction of the capacitor the open loop gain curve starts falling off at a rate of $-20\,\text{dB/decade}$ from its constant gain value at the new corner frequency f_x. When the frequency f_1 is reached, the rate of roll off becomes $-40\,\text{dB/decade}$ since, according to (3.9), the roll off rate of the compensated operational amplifier will be the sum of that of the uncompensated operational amplifier and the compensating network. The gain will decrease at a rate of $-60\,\text{dB/decade}$ from f_2 onwards.

In order to obtain a stable circuit having a maximum possible bandwidth we choose the value of C in such a way that the open loop gain curve of the compensated operational amplifier meets the closed loop gain curve with rates of closure of $20\,\text{dB/decade}$ on one side of the intersection and $40\,\text{dB/decade}$ on the other side to provide a phase margin of at least $45°$. If the pole was chosen to occur at a lower frequency than f_x then we would have had a more stable circuit with larger phase margin, but with a smaller bandwidth. If it was chosen to occur at a higher frequency than f_x the circuit would have then stayed unstable.

Worked example 3.6
The operational amplifier used in the circuit of Fig. 3.3(a) has a low-frequency differential mode gain of 80 dB and corner frequencies at 20 kHz and 200 kHz. Examine the stability of the circuit and if necessary apply lag frequency compensation, given that $R_1 = 1\,\text{k}\Omega$, $R_2 = 100\,\text{k}\Omega$ and the output resistance of the preceding stage at the compensating point is $10\,\text{k}\Omega$.

Solution
See Fig. ex.3.6 and 3.7.
The closed loop gain $\simeq \dfrac{1}{\beta} = 1 + \dfrac{Z_2}{Z_1} = 1 + 100 = 101$

$20\log\dfrac{1}{\beta} = 20\log 101 \simeq 40\,\text{dB}.$

The $\dfrac{1}{\beta}$ curve meets the A_{OL} curve with 40 dB/decade rate of closure on both sides of the crossing point. So the circuit is unstable. Choose f_x so that the second corner frequency of the compensated open loop gain curve sits on

Fig. ex.3.6/3.7

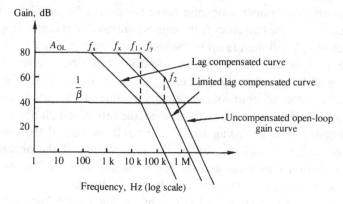

Frequency, Hz (log scale)

the $\dfrac{1}{\beta}$ curve thus giving a 45° phase margin. From the diagram, $f_x = 200\,\mathrm{Hz}$.
Again

$$f_x = \frac{1}{2\pi CR}$$

Therefore,

$$C = \frac{1}{2\pi f_x R} = \frac{1}{2\pi \times 200 \times 10 \times 10^3} = 79.6\,\mathrm{nF}$$

The above solution shows that the procedure for the lag frequency
compensation referring to Fig. 3.8 is as follows:

(i)   draw a vertical line at f_1.
(ii) from the crossing point of this line and the closed loop gain curve
      draw a line with a slope of $-20\,\mathrm{dB/decade}$ toward the left of the
      vertical line.
(iii) The new pole should be introduced at the frequency at which this
      line meets the uncompensated open loop gain curve.

Limited lag compensation
   If we add a resistor in series with the capacitor between point P and
ground as shown in Fig. 3.9 then a zero is introduced in addition to the pole
in the transfer function of the compensated operational amplifier open loop
gain curve. From Fig. 3.9 we observe that the new open loop gain of the
operational amplifier

$$A_{\mathrm{OL}} = A_{\mathrm{OL}}(\mathrm{old}) \times \frac{1 + j2\pi fCR_2}{1 + j2\pi fC(R + R_2)}$$

Fig. 3.8. Effect of introduction of a pole on the open loop gain curve of an operational amplifier.

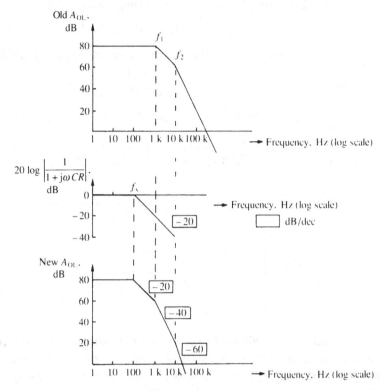

Fig. 3.9. Addition of a resistor in series with the capacitor between the compensating point and ground for limited lag frequency compensation.

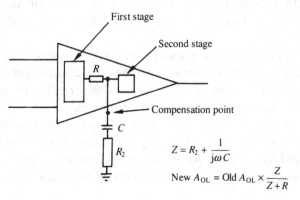

$$Z = R_2 + \frac{1}{j\omega C}$$

$$\text{New } A_{OL} = \text{Old } A_{OL} \times \frac{Z}{Z + R}$$

now taking logarithms on both sides we obtain

$$20 \log A_{OL} \, dB = 20 \log A_{OL}(old) + 20 \log(1 + j2\pi fCR_2)$$
$$- 20 \log(1 + j2\pi fC(R + R_2)) \, dB \qquad (3.10)$$

The second and third terms on the right hand side of (3.10) will introduce two corner frequencies one at $f_x = \dfrac{1}{2\pi C(R + R_2)}$ and the other at $f_y = \dfrac{1}{2\pi CR_2}$. The magnitude response will fall off at a rate of 20 dB/decade from f_x and from f_y onwards it would be constant since the $+20$ dB/decade rise due to the zero at f_y will cancel the -20 dB/decade fall.

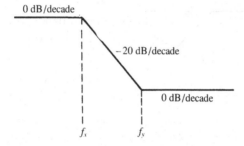

To use this method effectively, i.e., to have the optimum phase margin of $45°$, we should make the value of f_y the same as the first corner frequency of the uncompensated operational amplifier and select a value for f_x lower than the first corner frequency of the uncompensated operational amplifier. Figure 3.10 shows that the open loop gain curve of the compensated operational amplifier has a constant gain from d.c. to f_x, after which it falls off at a rate of -20 db/decade. The pole of the uncompensated operational amplifier at f_1 is effectively cancelled by the zero of the compensating network, thus letting the gain of the compensated operational amplifier continue to fall at the same rate of -20 dB/decade through the frequency f_1 up to the frequency f_2. From this frequency onwards the rate of roll off is -40 dB/decade.

The value of f_x should be chosen in such a way that the intersection of the closed loop gain curve of the amplifier and the compensated operational amplifier open loop gain curve occurs at the frequency, f_2. Thus the rate of closure between the two curves will be 20 dB/decade on one side and 40 dB/decade on the other side of the point of intersection providing a phase margin of $45°$, and a conditionally stable circuit with maximum possible bandwidth is obtained. Again as in the lag compensation method if the new pole was chosen to occur at a lower frequency than 'f_x' then we

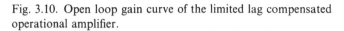

Fig. 3.10. Open loop gain curve of the limited lag compensated operational amplifier.

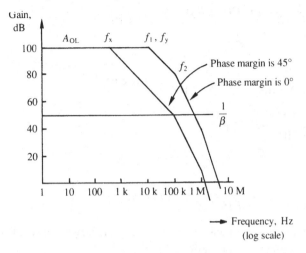

would have had a more stable circuit but with smaller bandwidth. If it was chosen to occur at a higher frequency than 'f_x' then we would have failed to stabilise the circuit. From the following Worked example 3.7 the reader can observe that the limited lag compensation gives a larger bandwidth than that which could be obtained by using the lag compensation.

Worked example 3.7

Repeat Ex.3.6. Apply limited lag frequency compensation if necessary. Comment on the useful bandwidth.

Solution

See Fig. ex.3.6 & 3.7.

In the solution of Ex. 3.6 it was found that the circuit needed to be compensated.

Apply limited lag compensation to the circuit by making $f_y = f_1$ (where pole-zero cancellation occurs) and choosing f_x so that the second corner frequency of the compensated open loop curve occurs at the crossing point of the closed loop gain curve and the new open loop gain curve thus giving a phase margin of 45°. From the diagram $f_x = 2\,\text{kHz}$ and $f_y = 20\,\text{kHz}$.

Again

$$f_y = \frac{1}{2\pi CR_2} \quad \text{or,} \quad CR_2 = \frac{1}{2\pi \times 20 \times 10^3}$$

Also

$$f_x = \frac{1}{2\pi C(R+R_1)} \quad \text{or,} \quad C(R+R_2) = \frac{1}{2\pi \times 2 \times 10^3}$$

Therefore,

$$CR = \frac{1}{2\pi \times 2 \times 10^3} - \frac{1}{2\pi \times 20 \times 10^3} = 71.6 \times 10^{-6}$$

$$\therefore \quad C = \frac{71.6 \times 10^{-6}}{10 \times 10^3} = 7.16\,\text{nF}$$

Now

$$R_2 = \frac{1}{2\pi \times 20 \times 10^3 \times 7.16 \times 10^{-9}} = 1112\,\Omega$$

In this case the limited lag compensation gives 1 decade larger bandwidth than that obtained by lag compensation.

Worked example 3.8
Examine the frequency stability of the operational amplifier circuit of Fig. ex.3.8 and if necessary apply limited lag frequency compensation. The open loop gain of the operational amplifier is 100 dB at low frequency and has corner frequencies at 1000 Hz and 100 000 Hz. A frequency compensation point is provided with a self resistance of 4 kΩ.

Solution
1st step: Draw operational amplifier open loop gain curve
2nd step: Calculate $1/\beta$

Fig. ex.3.8

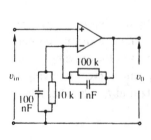

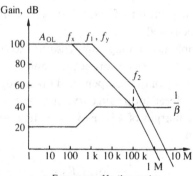

Frequency, Hz (log scale)

$$1/\beta = 1 + \frac{Z_2}{Z_1} = 1 + \cfrac{100 \times 10^3 \times \cfrac{1}{j2\pi f \times 1 \times 10^{-9}}}{100 \times 10^3 + \cfrac{1}{j2\pi f \times 1 \times 10^{-9}}}\Bigg/\cfrac{10 \times 10^3 \times \cfrac{1}{j2\pi f \times 100 \times 10^{-9}}}{10 \times 10^3 + \cfrac{1}{j2\pi f \times 100 \times 10^{-9}}}$$

$$= \frac{11(1+j0.00577f)}{(1+j0.000628f)}$$

So the frequency response of $1/\beta$ has:

a pole at $f_p = 1/0.000628 = 1592\,\text{Hz}$

a zero at $f_x = 1/0.00577 = 173.3\,\text{Hz}$

and

a low frequency gain of $20\log_{10} 11 = 20.8\,\text{dB}$

3rd step: Draw $1/\beta$
The open loop and closed loop gain curves meet at 40 dB/decade rate of closure on both sides of the intersection thus giving a phase margin of 0°. So the circuit is unstable.

4th step: Choose $f_y = f_1$ where pole-zero cancellation occurs. Choose f_x so that the new open loop gain curve meets the $1/\beta$ curve at f_2. From the diagram we see that the phase margin is now 45° and $f_x = 100\,\text{Hz}, f_y = 1\,\text{kHz}$.

$$f_y = \frac{1}{2\pi CR_2} \quad \text{or} \quad CR_2 = \frac{1}{2\pi \times 1000}$$

Again

$$f_x = \frac{1}{2\pi C(R+R_2)} \quad \text{or} \quad C(R+R_2) = \frac{1}{2\pi \times 100}$$

Therefore

$$CR = \frac{1}{2\pi \times 100} \cdot \frac{9}{10}$$

or

$$C = \frac{9}{2\pi \times 1000 \times 4 \times 10^3} = 360\,\text{nF}$$

and

$$R_2 = \frac{1}{2\pi \times 1000 \times 0.36 \times 10^{-6}} = 442\,\Omega.$$

3.2   Applications

3.2.1   Instrumentation amplifiers

The main use of this type of amplifier is in instrumentation systems where it amplifies small differential voltages riding on large unwanted common-mode voltages. The differential-mode gain of an instrumentation amplifier is often high, whereas, its common-mode gain is very low, thus giving a very high common-mode rejection ratio. The input signal to the instrumentation amplifier is usually the output voltage of a bridge circuit with a variable resistance transducer in one of its arms as shown in Fig. 3.11. The bridge has to be provided with an excitation voltage. With a constant excitation voltage V applied across A and B, the differential or error voltage ΔV across C and D will be

$$\Delta V = \Delta R/4R \left[\frac{1}{1 + \frac{\Delta R}{2R}} \right] V$$

If we can make $2R \gg \Delta R$, then

$$\Delta V \simeq \Delta R/4R \cdot V \tag{3.11}$$

Strain gauges, thermistors and photo-resistors are variable resistance transducers. Active transducers such as thermocouples and piezoelectric devices which generate voltage can be connected to the instrumentation amplifier directly. The presence of a common-mode voltage and a

Fig. 3.11. A bridge circuit.

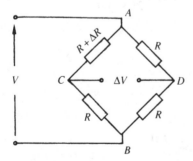

Fig. 3.12. An instrumentation amplifier with high input impedances and differential gain.

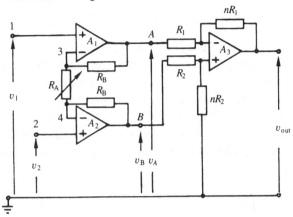

differential voltage is the characteristic of most transducers. The common-mode voltage can be a d.c. level or a noise pick-up.

The circuit of an instrumentation amplifier is shown in Fig. 3.12. The input pair of the operational amplifiers A_1 and A_2 provide high impedance at both inputs and high differential gain. By matching input pairs, very low drifts can be achieved. The third operational amplifier, A_3 at the output stage converts the differential signal from the input pair into a single-ended output signal. It also provides a low output impedance. The gain can be changed by the variable resistor R_A.

Analysis

To analyse the circuit we will make use of the superposition theorem. Let us first consider the output stage. If we assume that the voltage applied to terminal 'B' is zero, then the terminal can be effectively grounded and the operational amplifier A_3 will be connected in an inverting mode. Therefore, if a voltage v_A is applied to terminal 'A' the output voltage will be

$$v'_{out} = -nv_A$$

Now if we make $v_A = 0$, terminal 'A' will be at ground potential and the amplifier will be in the noninverting configuration. In this case the output voltage will be

$$v''_{out} = \frac{nR_2}{R_2 + nR_2}\left(1 + \frac{nR_1}{R_1}\right)v_B = nv_B$$

Using the superposition theorem the output voltage can be written in terms of v_A and v_B

$$v_{\text{out}} = v'_{\text{out}} + v''_{\text{out}} = n(v_B - v_A) \tag{3.12}$$

Now let us consider the input stage of the instrumentation amplifier. When the input voltage v_2 is zero, it can be assumed that terminal '2' is connected to ground. Then the operational amplifier A_1 acts in the noninverting mode, whereas, the other amplifier A_2 works as an inverting amplifier to the input signal v_1 since the differential-mode voltage of the operational amplifiers is zero, and the potential at nodes '3' and '4' are the same as those at '1' and '2' respectively. Thus the output voltages at terminals 'A' and 'B' are respectively given by

$$v'_A = \left(1 + \frac{R_B}{R_A}\right) v_1$$

and

$$v'_B = -v_1 \cdot \frac{R_B}{R_A}$$

Using a similar argument it can be said that when $v_1 = 0$, the operational amplifiers A_1 and A_2 act in the inverting and noninverting modes respectively to the input voltage v_2. Thus in this case, the output voltages at terminals 'A' and 'B' are

$$v''_A = -v_2 \cdot \frac{R_B}{R_A}$$

and

$$v''_B = \left(1 + \frac{R_B}{R_A}\right) \cdot v_2$$

respectively. Using the superposition theorem again we find the output voltages at terminals 'A' and 'B' for input voltages v_1 and v_2 to be

$$v_A = v'_A + v''_A = \left(1 + \frac{R_B}{R_A}\right) v_1 - \frac{R_B}{R_A} \cdot v_2 \tag{3.13}$$

and

$$v_B = v'_B + v''_B = \left(-\frac{R_B}{R_A}\right) v_1 + \left(1 + \frac{R_B}{R_A}\right) v_2 \tag{3.14}$$

From (3.12), (3.13) and (3.14) we obtain the relationship between the differential input voltage and the output voltage

Fig. 3.13. The schematic of an integrated circuit instrumentation amplifier.

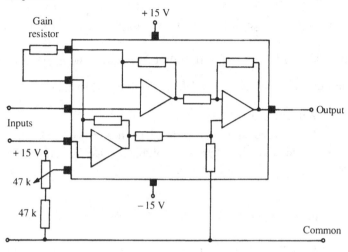

$$v_{\text{out}} = -n\left(1 + \frac{2R_B}{R_A}\right)(v_1 - v_2)$$

or

$$v_{\text{out}} = -n\left(1 + \frac{2R_B}{R_A}\right)\Delta v \qquad (3.15)$$

Therefore the gain of the instrumentation amplifier is given by $-n\left(1 + \frac{2R_B}{R_A}\right)$. The negative sign indicates that the output is phase shifted by 180°. By adjusting the variable resistor R_A the gain can be changed over a wide range. The component values are usually chosen so that the gain is shared roughly equally by the two stages of the amplifier.

Figure 3.13 shows the schematic of an integrated circuit instrumentation amplifier. These are generally balanced to produce zero output voltage for zero input voltage when installed, or perhaps before a test run, but due to changes in input offset current and voltage with time, temperature and power supply variation and also due to any change in output noise voltage, some offset error will be produced at the output. The total offset error voltage can be calculated using the following equation,

$$E_{\text{os}} = \pm \frac{\Delta E_{\text{os}}}{\Delta T} \cdot \Delta T \pm \frac{\Delta E_{\text{os}}}{\Delta V_s} \cdot \Delta V_s \pm \frac{\Delta E_{\text{os}}}{\Delta t} \cdot \Delta t$$

$$\pm R_s \cdot \frac{\Delta I_{\text{os}}}{\Delta T} \cdot \Delta T \pm \Delta E_n \qquad (3.16)$$

where $\dfrac{\Delta E_{os}}{\Delta T}$ is the output offset voltage drift with temperature, $\dfrac{\Delta E_{os}}{\Delta V_s}$ is the output offset voltage drift with power supply, $\dfrac{\Delta E_{os}}{\Delta t}$ is the output offset voltage drift with time, $\dfrac{\Delta I_{os}}{\Delta T}$ is the offset current drift with temperature, ΔE_n is the output noise voltage and R_s is the source resistance.

Worked example 3.9

The bridge shown in Fig. 3.11 has $V = 6\,V$, $R = 8.2\,k\Omega$ and $\Delta R = 60\,\Omega$. Calculate the differential voltage ΔV. Design a suitable circuit to amplify this voltage by a factor of 1200.

Solution

From (3.11)

$$\Delta V = \frac{\Delta R}{4R}\left(\frac{1}{1 + \dfrac{\Delta R}{2R}}\right)V$$

or

$$\Delta V = \frac{60}{4 \times 8.2 \times 10^3}\left(\frac{1}{1 + \dfrac{60}{2 \times 8.2 \times 10^3}}\right) \times 6 = 10.94\,mV$$

Again from (3.15), the overall gain

$$n\left(1 + \frac{2R_B}{R_A}\right) = 1200$$

say

$$n = 30, \text{ then } 1 + \frac{2R_B}{R_A} = 40$$

Choose $R_1 = R_2 = R_A = 1\,k\Omega$. Therefore

$$R_B = 19.5\,k\Omega$$

Worked example 3.10

An integrated circuit instrumentation amplifier has the following specifications at a gain of 1000:

$$\frac{\Delta E_{os}}{\Delta T} = \pm 10\,mV/°C, \frac{\Delta E_{os}}{\Delta V_s} = \pm 50\,mV/V, \frac{\Delta E_{os}}{\Delta t} = \pm 5\,mV/24\,h,$$

$$\frac{\Delta I_{os}}{\Delta T} = \pm 1\,nA/°C \text{ and } \Delta E_n = 6\,mV\,rms$$

If this amplifier is balanced to zero output for zero input before being connected to the bridge described in Ex. 3.9 what would be the offset error voltage after a week if the temperature change during this period is 19°C and the variation in the power supply is 5%. The power supply needed for the amplifier is $\pm 15\,\mathrm{V}$.

Solution

$$R_s = (4.1 + 4.1) \times 10^3 = 8.2\,\mathrm{k\Omega}$$

Substituting the given data in (3.16)

$$E_{os} = \pm (10 \times 10^{-3} \times 19) \pm (50 \times 10^{-3} \times 15 \times 0.05) \pm (5 \times 10^{-3} \times 7)$$
$$\pm (8.2 \times 10^3 \times 1 \times 10^{-9} \times 19)$$
$$\pm (6 \times 10^{-3} \times 1.414)$$
$$= 271.2\,\mathrm{mV}$$

The calculated offset error of $\pm 271.2\,\mathrm{mV}$ is a worst case maximum value assuming all terms are additive.

3.2.2 Comparators

A comparator as its name implies compares the amplitude of a varying signal either with that of another varying signal or with that of a reference voltage. It is widely used in electronic systems to sense when an input signal reaches some threshold value. By limiting the output voltage to suitable levels it is possible for a comparator to drive digital logic circuits directly. Figure 3.14 shows that an operational amplifier can be used without any external circuitry to perform the function of a comparator. In this case the input signal is applied to the inverting terminal of the amplifier and a reference voltage to the noninverting terminal. The amplifier output

Fig. 3.14. An operational amplifier as a basic comparator.

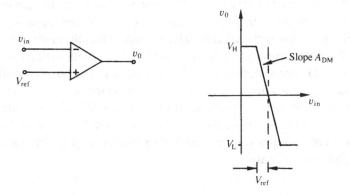

Fig. 3.15(a). The regenerative comparator circuit and (b) its input/output characteristic.

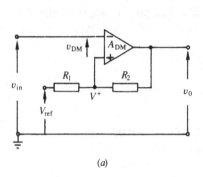

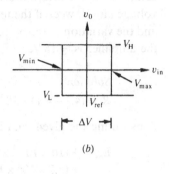

(b)

(a)

is V_H if the signal exceeds the magnitude of the reference voltage by at least 1 mV typically, in other words $v_{in} - V_{ref}$ is greater than 1 mV. Now if the magnitude of the input signal is reduced gradually then at a certain instant $v_{in} - V_{ref}$ becomes less than -1 mV and the output of the amplifier becomes V_L. Again the amplifier output changes state (V_H) when the signal exceeds the magnitude of the reference voltage. It is apparent from the figure that the input should vary by an amount equal to $(V_H - V_L)/A_{DM}$, i.e., 2 mV in this particular case, before any change of state occurs at the output. Therefore for a slowly varying input signal it is necessary to speed up the transition of the output.

This can be achieved with the help of a positive feedback circuit as shown in Fig. 3.15(a). In this circuit, known as the regenerative comparator or Schmitt trigger, positive feedback is applied via the potential divider R_1 and R_2. The input signal is applied to the inverting terminal and the reference voltage to the noninverting terminal via the resistor R_1. Let us assume that initially the input signal v_{in} is much less than the reference voltage, V_{ref}, then the differential mode voltage v_{DM} will be negative and since the operational amplifier is connected in the inverting mode the output will be equal to the positive saturation voltage V_H. Now let us suppose that the input voltage v_{in} rises. When it becomes equal to the potential at the noninverting terminal, $V^+ = V_{max}$ say, the output voltage V_H falls slightly. The positive feedback increases v_{DM} which in turn accelerates the drop in the output voltage until the latter reaches the negative saturation voltage V_L. The transition is regenerative and takes typically less than 1 μs. It is independent of the rate of change of the input voltage. Any further increment in v_{in} will not cause any change in the output.

Now if the input voltage v_{in} is decreased, the output voltage will stay at V_L until v_{in} falls to the new value of $V^+ = V_{min}$, then the positive feedback will force the output to rise to V_H. Again this transition is both rapid and regenerative. Any further reduction of the input voltage will not change the output voltage.

The upper threshold voltage, when the input is rising, is given by

$$V_{max} = V_H - \frac{V_H - V_{ref}}{R_1 + R_2} \cdot R_2 = \frac{V_{ref} \cdot R_2 + V_H \cdot R_1}{R_1 + R_2} \qquad (3.17)$$

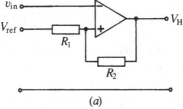

(a)

Similarly the lower threshold voltage, when the output is falling, is given by

$$V_{min} = V_L - \frac{V_L - V_{ref}}{R_1 + R_2} \cdot R_2 = \frac{V_{ref} \cdot R_2 + V_L \cdot R_1}{R_1 + R_2} \qquad (3.18)$$

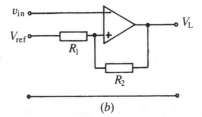

(b)

Therefore the transition takes place at different values of the input signal depending on whether it is increasing or decreasing. The difference between V_{max} and V_{min} is the amount of hysteresis and is given by

$$\Delta V = V_{max} - V_{min} = R_1/(R_1 + R_2) \cdot (V_H - V_L) \qquad (3.19)$$

From the above expression we may note that the amount of hysteresis is proportional to the feedback factor. Figure 3.15(b) shows the variation in the output voltage of the comparator with the variation in the input voltage.

The upper and lower output voltage levels are often set by the power supply voltages, they are typically 1 volt less than the power supply voltages due to various voltage drops in the output circuit of the operational

Fig. 3.16. A regenerative comparator circuit with output voltage levels at −4.6 volts and 3.6 volts.

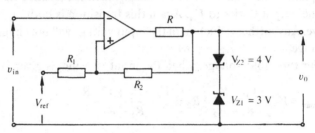

amplifier. However, for the use of comparators in conjunction with some digital logic circuits the upper output voltage level must have a value in the range of 2.4–5 volts and the lower output voltage level a value around zero. Zener diodes are often used for this purpose as shown in Fig. 3.16. Whatever the power supply voltages to the circuit are, the output voltage levels are determined by the Zener voltages of the diodes neglecting the voltage drop across the forward conducting Zener diode. The value of resistor R is so chosen that the amplifier output current is limited to the recommended bias current for the Zener diodes. In the circuit when the input voltage v_{in} is smaller than the reference voltage V_{ref}, the output of the operational amplifier assumes the positive state. Then the diode Z_1 becomes reverse biased and the diode Z_2 goes forward biased. Hence the output voltage becomes

V_{out} = reverse voltage drop of Z_1 + forward voltage drop of Z_2

$= 3 + 0.6 = 3.6$ volts

Again when the input voltage is greater than the reference voltage V_{ref}, the output assumes the negative state. Now the Zener diode Z_1 becomes forward biased and the diode Z_2 goes reverse biased. The output voltage will then become

$V_{out} = -$ (forward voltage drop of Z_1 + reverse voltage drop of Z_2)

$= -(0.6 + 4) = -4.6$ volts

This comparator circuit converts a sine wave or triangular wave input to a square wave output and hence is also referred to as a squaring circuit.

Worked example 3.11
    In the circuit of Fig. 3.15(a) $R_1 = 1$ kΩ. Choose the values of R_2 and the reference voltage, V_{ref} so that the maximum and minimum input voltages required to switch over the output voltage levels of the circuit are 4.7 V and 1.5 V respectively. Given $V_H = 12$ V and $V_L = -12$ V.

Solution

Substituting 4.7 V, 12 V and 1 kΩ for V_{max}, V_H and R_1 respectively in the expression (3.17) we get

$$4.7 = \frac{V_{ref} \cdot R_2 + 12 \times 1 \times 10^3}{R_2 + 1 \times 10^3}$$

or

$$V_{ref}R_2 - 4.7R_2 = 4.7 \times 10^3 - 12 \times 10^3$$

Similarly substituting 1.5 V, -12 V and 1 kΩ for V_{min}, V_L and R_1 respectively in the expression (3.18) we get

$$1.5 = \frac{V_{ref} \cdot R_2 - 12 \times 1 \times 10^3}{R_2 + 1 \times 10^3}$$

or

$$V_{ref}R_2 - 1.5R_2 = 1.5 \times 10^3 + 12 \times 10^3$$

Solving the above two expressions

$$\begin{array}{r} V_{ref}R_2 - 4.7R_2 = -7.3 \times 10^3 \\ -V_{ref}R_2 + 1.5R_2 = -13.5 \times 10^3 \\ \hline 3.2R_2 = 20.8 \times 10^3 \end{array}$$

$$\therefore \quad R_2 = 6.5 \,\text{k}\Omega$$

$$V_{ref} = \frac{-7.3 \times 10^3 + 4.7 \times 6.5 \times 10^3}{6.5 \times 10^3} = 3.58 \,\text{V}$$

Worked example 3.12

The operational amplifier used in the circuit of Fig. 3.15(a) has a slew rate of 1.5 V/μs. Draw the output voltage waveform when a triangular wave of ± 5 V amplitude and frequency 10 kHz is applied to the input. Given $R_1 = 1 \,\text{k}\Omega$, $R_2 = 10 \,\text{k}\Omega$, $V_{ref} = 2$ V, $V_H = 9$ V and $V_L = -9$ V.

Solution

$$V_{max} = \frac{V_{ref}R_2 + V_H R_1}{R_1 + R_2} = \frac{2 \times 10 \times 10^3 + 9 \times 1 \times 10^3}{(1 + 10)10^3} = 2.64 \,\text{V}$$

$$V_{min} = \frac{V_{ref}R_2 + V_L R_1}{R_1 + R_2} = \frac{2 \times 10 \times 10^3 - 9 \times 1 \times 10^3}{(1 + 10)10^3} = 1 \,\text{V}$$

1.5 V change in the output takes 1 μs

18 V change in the output takes $\dfrac{18}{1.5} = 12 \,\mu s$.

Fig. ex.3.12

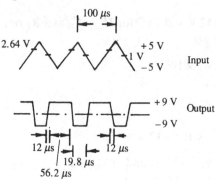

3.2.3 Precision rectifiers

Semiconductor diodes are used in rectifier circuits in order to convert a.c. voltages to pulsating d.c. voltages and allow current to flow in one direction only. A conventional rectifier circuit, as shown in Fig. 3.17(a), however, does not rectify very small signals and also introduces errors into large input signals due to the nonlinear characteristics of the diodes. To overcome these problems an operational amplifier can be used in conjunction with diodes and resistors as shown in Fig. 3.17(b). This circuit works as a half wave rectifier and to understand the principle of operation of the circuit we first consider the positive half of an alternating voltage applied to the input terminal. Since the operational amplifier is connected in the inverting mode, its output v_A at terminal 'A' will be negative. The diode D_1 will be forward biased but the diode D_2 will be

Fig. 3.17(a). A conventional half wave rectifier circuit.
(b). A precision half wave rectifier circuit.

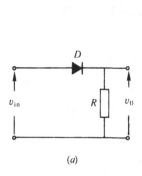

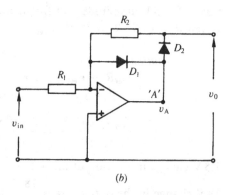

(a)   (b)

Fig. 3.18(a). Analysis of the circuit of Fig. 3.17(b) for positive input voltage.

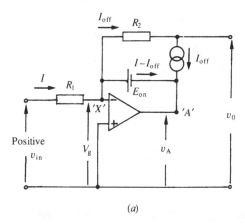

(a)

reverse biased. The circuit can be redrawn as shown in Fig. 3.18(a) where the diode D_1 has been replaced by a d.c. source, 'E_{ON}' of magnitude equal to the forward voltage drop of a diode, say 0.6 volt for a silicon diode, and D_2 has been replaced by a current source having a magnitude equal to the leakage current flowing through D_2 due to reverse biasing. Now assuming that the operational amplifier behaves ideally, it can be said that the terminal 'X' is grounded, therefore the output voltage v_0 is equal to the voltage drop across the resistor R_2 due to the leakage current I_{OFF} flowing through it. The rest of the current I will flow to the terminal 'A' via the d.c. source 'E_{ON}'. Thus we may write

$$v_0 = -R_2 I_{OFF} \qquad (3.20)$$

and

$$v_A = V_f$$

where V_f is the forward voltage drop of a silicon diode.

Let us now consider the negative half of the alternating input voltage. The voltage v_A at terminal 'A' will be positive due to the inverting mode. The diode D_2 will be forward biased and diode D_1 will be reverse biased. The circuit can be represented as shown in Fig. 3.18(b) by replacing diodes D_1 and D_2 by a current generator and a d.c. source respectively. The output voltage in this case will be

$$v_0 = I'R_2 = (v_{in}/R_1 - I_{OFF})R_2 \text{ where } I' = I - I_{OFF}$$

$$= v_{in}R_2/R_1 - I_{OFF}R_2 \qquad (3.21)$$

Fig. 3.18(b). Analysis of the circuit of Fig. 3.17(b) for negative input voltage.

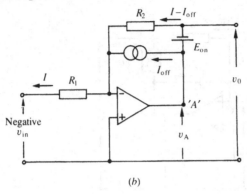

(b)

Fig. 3.19(a). Input/output characteristic and the input and output voltage waveforms for the precision half wave rectifier.

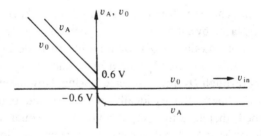

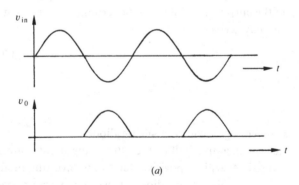

(a)

and

$$v_A = v_o + V_f$$

Figure 3.19(a) shows the variation of v_o and v_A with v_{in}. A typical value of I_{OFF} is 10 nA which if R_2 is 1 kΩ would give,

$$I_{OFF}R_2 = 10\,\mu V$$

Fig. 3.19(b). Input/output characteristic and the input and output voltage waveforms for a precision half wave rectifier with diodes connected in opposite directions to those in the circuit of Fig. 3.17(b).

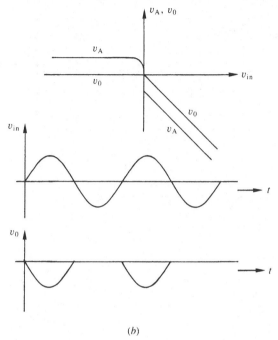

(b)

So we may neglect the term $I_{OFF}R_2$ in (3.20) and (3.21) provided $v_{in} \gg 10\,\mu V$, and the resistors R_1 and R_2 are at least of the same order. Thus the output voltage varies linearly with the input voltage when the latter is negative. The output is approximately zero for positive input voltages.

By changing the direction of the diodes in the circuit of Fig. 3.17(b) the output voltage can be made to vary with the input voltage in the manner as shown in Fig. 3.19(b).

The full wave rectification can be obtained by using the half wave rectifier described above in conjunction with a summer as shown in Fig. 3.20(a). The gain of the half wave rectifier is unity, and the second operational amplifier adds the output of the half wave rectifier, v_2 to the input v_{in}. We can readily see from the figure that for a positive value of v_{in}, $v_2 = 0$, hence the output voltage is

$$v_o = -v_{in} \tag{3.22}$$

When the input voltage is negative the output voltage is

$$v_o = -[v_{in} + 2(-v_{in})] = v_{in} \tag{3.23}$$

Fig. 3.20(a). A precision full wave rectifier, (b) its input/output
characteristic and (c) its input and output voltage waveforms.

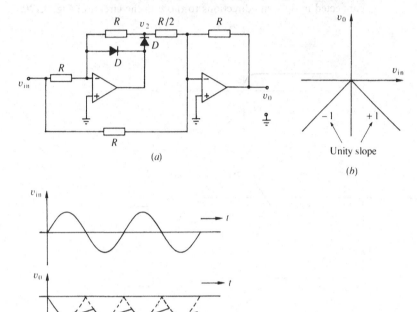

(a)

(b)

(c)

The input–output characteristic of the full wave rectifier is shown in Fig.
3.20(b). The output can be smoothed out with the help of a capacitor C
connected across the feedback resistor of the second operational amplifier
circuit. The input–output waveforms are shown in Fig. 3.20(c). This circuit
is widely used in digital voltmeters.

Worked example 3.13
In the circuit of Fig. 3.17(b) the silicon diodes have leakage
currents of $10\,\mu\text{A}$, $R_1 = 1\,\text{k}\Omega$ and $R_2 = 4.7\,\text{k}\Omega$. Draw the output voltage
waveform for a 50 Hz sinusoidal input of $2\,\text{V}_{\text{PTP}}$ amplitude.

Solution
When v_{in} is positive,

$$v_{\text{out}} = -R_2 I_{\text{off}} = -4.7 \times 10^3 \times 10 \times 10^{-6}$$

$$= -0.047\,\text{V}$$

Fig. ex.3.13

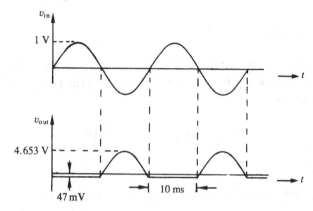

When v_{in} is negative,

$$v_{out} = v_{in} \times \frac{4.7 \times 10^3}{1 \times 10^3} - 4.7 \times 10^3 \times 10 \times 10^{-6}$$

$$= 4.7v_{in} - 0.047 \text{ V}$$

Worked example 3.14

Design a full wave rectifier to deliver 15 V d.c. with 0.2 V_{PTP} ripple into a resistive load of 3 kΩ. The input is the same as that given in Ex. 3.13.

Solution

$$i = C\frac{dv}{dt}$$

or

$$\frac{i}{C}dt = dv$$

or

$$\frac{i}{C} \cdot \frac{1}{2f} = 0.2$$

Fig. ex.3.14

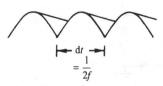

Therefore

$$C = \frac{15}{3 \times 10^3 \times 2 \times 50 \times 0.2} = 250 \, \mu\text{F}$$

Let us consider the full rectifier circuit of Fig. 3.20 and choose $R = 1 \, \text{k}\Omega$, then by using (3.21) and putting $R_1 = R_2 = R$, the output of the first stage (half rectifier) is given by

$$v_2 = I'R = \left(\frac{v_{in}}{R} - I_{off} \right) R = \left(\frac{1}{1 \times 10^3} - 10 \times 10^{-6} \right) 10^3$$

$$= 0.99 \, \text{V}$$

Therefore the value of the feedback resistor of the second stage (summer) is

$$15 \times 1 \times 10^3 = 15 \, \text{k}\Omega$$

and the resistor $R/2$ between the output of the first stage and the input of the second stage is

$$\frac{10^3}{2} = 500 \, \Omega$$

3.2.4   Logarithmic amplifiers

Logarithmic amplifiers are nonlinear circuits and may be used to provide very small input signals with large amplification and large input signals with relatively small amplification. Logarithm circuits together with antilogarithm circuits can perform analogue multiplication and division.

Logarithmic amplifiers use a diode or a transistor in the feedback path of an operational amplifier circuit as shown in Fig. 3.21. The output voltage of

Fig. 3.21. A logarithmic amplifier using a diode in the feedback path of an operational amplifier circuit.

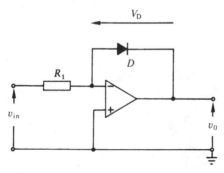

the circuit is proportional to the logarithm of the input voltage. The current–voltage relationship for a diode is given by

$$I_D = I_S(e^{qV_D/kT} - 1) \tag{3.24}$$

where

I_S = reverse saturation current
I_D = current through the diode
V_D = voltage across the diode
q = electronic charge 1.6×10^{-19} C
k = Boltzmann's constant 1.38×10^{-23} J/K
T = Temperature in K

If we assume that V_D is greater than 100 mV then

$e^{qV_D/kT} \gg 1$ at room temperature*

Therefore (3.24) can be written in the following form

$$I_D = I_S e^{qV_D/kT}\dagger \tag{3.25}$$

Taking logarithms of both sides,

$$\ln I_D/I_S = qV_D/kT$$

or

$$V_D = \frac{kT}{q} \ln I_D/I_S \tag{3.26}$$

Now if we consider the operational amplifier to be ideal then its differential mode voltage is equal to zero and therefore the output voltage v_o is equal but opposite to the voltage drop across the diode. Thus

$$v_o = -\frac{kT}{q} \ln I_D/I_S \tag{3.27}$$

Again

$$I_D = v_{in}/R_1 \tag{3.28}$$

since no current flows into the input terminal due to the infinite input impedance of the operational amplifier.

---

*   Thermal voltage $\dfrac{kT}{q} = 26$ mV at $T = 300$ K.

†   Here v_{in} is positive, otherwise the diode is reverse biased and $I_D = -I_S$ and v_o is negative.

Therefore we may write from (3.27) and (3.28),

$$v_o = -kT/q \ln[v_{in}/I_S R_1]$$

$$= -[kT/q \ln v_{in} - kT/q \ln I_S R_1] \qquad (3.29)$$

The second term on the right hand side of (3.29) is a constant so the log plot of the output voltage versus the $\log v_{in}$ is a straight line with a slope of $-2.3 \, kT/q$ mV/decade.

Connecting the diode in the reverse direction in the circuit of Fig. 3.21, a positive output voltage can be obtained from a negative input signal.

An n-p-n transistor connected in a common base configuration can replace the diode in the circuit of Fig. 3.21. If a positive voltage is applied to the input then the output will be negative since the operational amplifier is connected in the inverting mode. Thus the collector–base junction is zero biased and the emitter–base junction is forward biased, i.e., the transistor is in the forward active region and $I_C \simeq I_E$. The collector current, therefore, for a common-base transistor is

$$I_C \simeq I_{ES}(e^{q V_{EB}/kT} - 1) \qquad (3.30)$$

Again using the same argument as before, i.e., $e^{q V_{EB}/kT} \gg 1$ we obtain,

$$I_C \simeq I_{ES} e^{q V_{EB}/kT}$$

where

$$V_{EB} = \text{emitter–base voltage}$$

and

$$I_{ES} = \text{emitter–base current under small reverse bias}$$

Since the base of the transistor is grounded, the output voltage of the amplifier v_o is equal to the emitter–base voltage and we may write

$$v_o = -kT/q \ln \frac{I_C}{I_{ES}} = -kT/q \ln \frac{v_{in}}{R_1 I_{ES}} \qquad (3.31)$$

The output of a logarithmic amplifier is either positive or negative depending on the type of bipolar junction transistors used. In order to obtain a positive output from a negative input signal a p-n-p transistor has to be employed. The slope of $+2.3 kT/q$ gives an output voltage variation of approximately 60 mV per decade of the input voltage change at room temperature.

The circuits described above are simple but are limited in performance due to the linear temperature dependance of the thermal voltage, kT/q and

Fig. 3.22. A precision logarithmic amplifier giving an output voltage variation of -1 volt per decade of the input voltage.

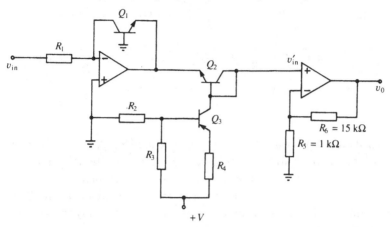

nonlinear temperature dependence of the quantities I_S and I_{ES}. The circuit shown in Fig. 3.22 uses a transistor Q_2 connected as a diode at the output of the first operational amplifier for temperature compensation. Ideally the constant current source Q_3 sets the input current at which the output voltage is zero. In order to obtain an output voltage variation of -1 volt per decade of the input voltage a noninverting amplifier is used with a voltage gain of $1\,\text{V}/60\,\text{mV} = 16$. A temperature sensitive resistor (thermistor) may be used in series with a resistor, instead of the resistor R_5 to compensate for variations in the thermal voltage kT/q.

The input voltage to the noninverting amplifier v'_{in} is equal to the difference between the emitter–base voltages of the transistors Q_1 and Q_2 and we may write

$$v'_{in} = v_{EB1} - v_{EB2} = -\left[\frac{kT}{q}\ln\frac{v_{in}/R_1}{I_{C2}}\right] = -\frac{kT}{q}\ln\frac{v_{in}}{R_1 I_{C2}} \qquad (3.32)$$

By choosing suitable values for resistors R_2, R_3 and R_4 we can make the output of the constant current source I_{C2} equal to $\dfrac{1}{R_1}$ amps.* Thus (3.32) will become

$$v'_{in} = -kT/q\ln v_{in}$$

* $I_{C2} \simeq I_{C3}$ since the bias current to the operational amplifier $\simeq 0$.

or

$$v_{out} = \left(1 + \frac{R_6}{R_5}\right)v'_{in} = -\left(1 + \frac{R_6}{R_5}\right)\frac{kT}{q} \cdot \ln v_{in}$$

$$= -2.3\frac{kT}{q}\left(1 + \frac{R_6}{R_5}\right)\log v_{in} \qquad (3.33)$$

If $R_6 = 15\,k\Omega$ and $R_5 = 1\,k\Omega$ then,

$$v_{out} \simeq -\log v_{in}$$

A matched monolithic transistor pair should be used for Q_1 and Q_2. Also in order to obtain an accurate logarithmic operation, operational amplifiers with low bias currents and offset voltages should be chosen.

By interchanging the diode and the resistor in the circuit of Fig. 3.21 a simple antilogarithm circuit can be obtained. It can be easily shown that the output voltage in this case will be

$$v_{out} = -RI_s \,\text{antilog}\,(qv_{in}/kT) \qquad (3.34)$$

Again a transistor can be used instead of the diode in the antilogarithm circuit. As in the case of logarithmic amplifiers these circuits are also unidirectional.

Worked example 3.15
Calculate the output voltage of the logarithmic amplifier circuit of Fig. 3.21 if the input voltage is 100 mV. Given $R_1 = 100\,k\Omega$, $I_{ES} = 1\,pA$ at an ambient temperature of 22 °C. What would be the output if the input is increased to 10 V?

Solution
From the expression (3.31), for $v_{in} = 100\,mV$,

$$v_{out} = -\frac{kT}{q}\ln\frac{v_{in}}{R_1 I_{ES}}$$

$$= -\frac{1.38 \times 10^{-23} \times 295}{1.6 \times 10^{-19}} \times \ln\frac{100 \times 10^{-3}}{100 \times 10^3 \times 1 \times 10^{-12}}$$

$$= -351\,mV$$

for $v_{in} = 10\,V$

$$v_{out} = -\frac{1.38 \times 10^{-23} \times 295}{1.6 \times 10^{-19}} \times \ln\frac{10}{100 \times 10^3 \times 1 \times 10^{-12}}$$

$$= -468\,mV$$

So the increment is 117 mV for 2 decades.

∴ increment per decade is approximately 60 mV.

Worked example 3.16

In the circuit of Fig. 3.22, $R_5 = 1\,\text{k}\Omega$ and $R_6 = 27\,\text{k}\Omega$. What would be the change in the output voltage per decade change in the input voltage at an ambient temperature of $55\,^\circ\text{C}$?

Solution

Let the input voltage be initially $v_{\text{in}1}$, and the corresponding output voltage $v_{\text{out}1}$.

Using (3.33)

$$v_{\text{out}1} = -2.3\frac{kT}{q}\left(1 + \frac{R_6}{R_5}\right)\log v_{\text{in}1} \tag{1}$$

Now increase the input voltage by a factor of 10 to $v_{\text{in}2}$; then the corresponding output voltage becomes $v_{\text{out}2}$.

Again using (3.33)

$$v_{\text{out}2} = -2.3\frac{kT}{q}\left(1 + \frac{R_6}{R_5}\right)\log v_{\text{in}2} \tag{2}$$

Subtracting (1) from (2),

$$v_{\text{out}2} - v_{\text{out}1} = -\left[2.3\frac{kT}{q}\left(1 + \frac{R_6}{R_5}\right)\log\frac{v_{\text{in}2}}{v_{\text{in}1}}\right]$$

$$= -2.3 \times \frac{1.38 \times 10^{-23} \times 328}{1.6 \times 10^{-19}}\left(1 + \frac{27 \times 10^3}{1 \times 10^3}\right)\log 10$$

$$= -1.82\ \text{volts}$$

3.2.5 Analogue computation

At the beginning of this chapter it was mentioned that operational amplifiers were at first used in analogue computation. In this section we will discuss the principles of operation of analogue computers and some of their applications. Operational amplifiers are normally used in analogue computation as sign reversal amplifiers, integrators, summers and also multipliers. In addition to these main components, potentiometers are used for the multiplication of the coefficients. Although an analogue computer solves differential equations and sometimes a need may arise for the use of a differentiator, its use is avoided because it presents increased noise level, risk of overloading and instability. The basic building blocks of analogue computers are described in the following sections.

Fig. 3.23. (*a*) A summer circuit and (*b*) its symbol.

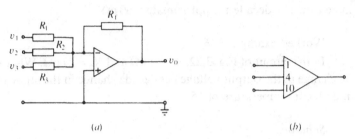

(*a*)   (*b*)

3.2.5.1 Summers

Figure 3.23(*a*) shows a typical circuit for a summer used in analogue computation. By assuming that the operational amplifier is ideal it can be shown easily that $v_0 = -(R_1/R_1 \cdot v_1 + R_1/R_2 \cdot v_2 + R_1/R_3 \cdot v_3)$. Therefore a number of voltages can be multiplied by different constants and added together, the constants being the ratio of the resistors. The symbol for a summer which avoids drawing all the resistors is shown in Fig. 3.23(*b*). The number beside each input terminal indicates the multiplication constant for that particular input. Any input not in use is grounded.

3.2.5.2 Integrators

The feedback component in an integrator is a capacitor. Figure 3.24(*a*) shows a typical integrator circuit used in analogue computation. Again assuming ideal performances for the operational amplifier we may write the output voltage v_0 in terms of the input voltages as follows,

$$v_0 = -1/C \int \left(\frac{v_1}{R_1} + \frac{v_2}{R_2} + \frac{v_3}{R_3} \right) dt \tag{3.35}$$

Thus a number of input voltages can be multiplied by different constants, then added together and integrated with respect to time. The constant of integration is the voltage at terminal '*A*' at $t = 0$. The potential divider formed by R_4 and R_5 charges the capacitor C and sets the initial conditions and then is disconnected as the integrator is set in the run mode. Often the feedback capacitor of the integrator is chosen to be $1 \mu F$ and the input resistors $1 M\Omega$ or $100 k\Omega$. The symbol for an integrator is shown in Fig. 3.24(*b*). The gain for each input is specified in the small rectangle and the initial condition, x_0, is shown fed into the triangle. The circle represents a resistive divider.

Fig. 3.24. (a) A typical integrator circuit and (b) its symbol.

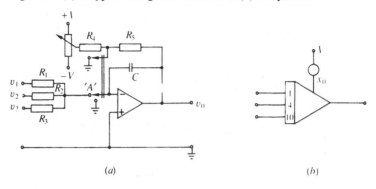

(a)   (b)

Fig. 3.25. The schematic of a multiplier circuit.

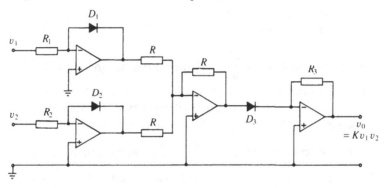

3.2.5.3 Multipliers

Figure 3.25 shows the schematic of a multiplier circuit built with two logarithm amplifiers and one antilogarithm amplifier. It is apparent from the figure that this circuit is capable of multiplying two voltages.

3.2.5.4 Applications

One of the main uses of an analogue computer is to solve differential equations which may simulate the behaviour of many mechanical systems. As an example let us find the displacement, x of a vibrating ball of mass, m on the end of a spring of stiffness, k in the presence of viscous damping described by the damping factor, f, from its equilibrium position. The equation of motion is given by

$$m\frac{d^2x}{dt^2} + f\frac{dx}{dt} + kx = 0$$

Fig. 3.26. Analogue computation circuit which solves the differential equation:

$$m\frac{d^2x}{dt^2} + f\frac{dx}{dt} + kx = 0.$$

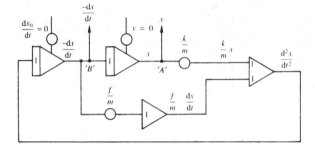

or

$$\frac{d^2x}{dt^2} = -\frac{f}{m}\frac{dx}{dt} - \frac{k}{m}x$$

We assume the value of $\frac{d^2x}{dt^2}$ and then using two integrators compute the value of x. The value of $-k/m \cdot x$ is found by using a summer to reverse the sign of x and a potentiometer to multiply it by k/m. Another potentiometer is used to multiply the output of the first integrator to obtain $-f/m \cdot dx/dt$. These two terms together yield $\frac{d^2x}{dt^2}$ which was assumed in the first place.

Figure 3.26 shows the circuit which will solve this equation.

Now if we set both integrators of the circuit at run mode simultaneously the voltages appearing at 'A' and 'B' will give the displacement and velocity of the ball respectively. These can be recorded on an oscilloscope or an XY-plotter.

Another application of analogue computation is to realise a class of filter circuit which will be described in the next section.

Worked example 3.17

Design an analogue computation circuit to solve the differential equatior

$$7\frac{d^2x}{dt^2} - 3\frac{dx}{dt} + 5x = 9$$

Solution

$$\frac{d^2x}{dt^2} = \frac{3}{7}\frac{dx}{dt} - \frac{5}{7}x + \frac{9}{7}$$

Fig. ex.3.17

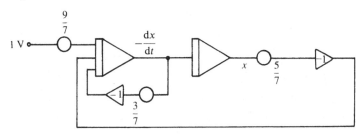

3.2.5.5 Scaling

Sometimes the magnitudes and rates of change with respect to time of the variables in the computer and the problem under study have real limitations. Hence in order to ensure accurate computer solutions, the corresponding variables in the computer and the system under study must be related by magnitude and time scale factors.

Magnitude scaling: In general we assume $s = a_m S$ where s is physical variable, S is computer variable and a_m is magnitude scale factor.

Thus we have in the equation to be simulated

$$\frac{ds}{dt} = a_s \frac{dS}{dt} \quad ; \quad \frac{d^n s}{dt^n} = a_s \frac{d^n S}{dt^n} \tag{3.36}$$

Time scaling: In order to transform from the real time, t, to computer time, T, we use the time scale factor a_t as follows,

$$t = a_t T$$

$$\frac{d}{dt} = \frac{1}{a_t} \cdot \frac{d}{dT} \quad ; \quad \frac{d^n}{dt^n} = \frac{1}{a_t^n} \cdot \frac{d}{dT^n} \tag{3.37}$$

The following worked example shows the application of time and magnitude scaling to a problem.

Worked example 3.18

A tank with a tap at its bottom is filled with water at a rate of $2\,\mathrm{m^3/s}$. The water flows out through the tap at a rate of $x\,\mathrm{m^3/s}$ where x is the height of the water level in the tank. If the area of the tank is $7200\,\mathrm{m^2}$, design an analogue computer to indicate the height of the water level in the tank.

Solution

The physical equation is

$$7200\frac{dx}{dt} + x = 2 \tag{1}$$

Fig. ex. 3.18(a)

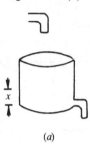

(a)

Substituting $\dfrac{dx}{dt} = 0$ into the above expression yields the maximum value of x,

$$x_m = 2\,\text{m}$$

Say the permissible computing voltage $= 10$ volts. Therefore using amplitude scaling

$$a_S = \frac{s}{S} = \frac{2}{10}\,\text{m/V}$$

Equation (1) becomes

$$7200 \times \frac{2}{10}\frac{dX}{dt} + \frac{2}{10}X = 2$$

or

$$7200\frac{dX}{dt} + X = 10 \qquad (2)$$

The initial rate of rise is obtained by putting $x = 0$.

$$\therefore\ 7200\frac{dx}{dt} = 2$$

or

$$\frac{dx}{dt} = \frac{2}{7200}$$

If this rate is maintained the final height (2 m) would be reached in 7200 s, i.e. 2 h. Let this be scaled to 10 s

$$\therefore\ a_t = \frac{7200}{10} = 720$$

Fig. ex. 3.18(b)

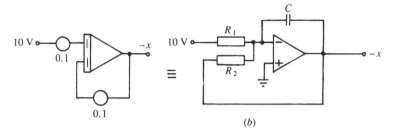

(b)

∴ equation (2) becomes

$$\frac{7200}{720}\frac{\mathrm{d}X}{\mathrm{d}T} + X = 10$$

or

$$\frac{\mathrm{d}X}{\mathrm{d}T} = -\frac{1}{10}X + 1$$

$$\frac{1}{R_1 C} = \frac{1}{R_2 C} = \frac{1}{10}$$

Choose $C = 1\,\mu\mathrm{F}$.

$$\therefore R_1 = R_2 = 10\,\mathrm{M}\Omega$$

3.2.6 Active filters

In low frequency filter circuits inductors are very bulky and their usage can be avoided by using operational amplifiers in conjunction with resistors and capacitors which can mimic the properties of inductors. Thus it is possible to build inductorless filters of various types in integrated circuit form with the ideal properties of RLC filters. In this section we will mainly consider low-pass and high-pass filters. The band-pass and band-reject filters can be easily constructed by cascading low-pass and high-pass filters. Figure 3.27 shows the ideal and practical gain responses of these filters. The Butterworth response which gives 'maximally flat' response in the pass band will be mostly used in this section. The rate of attenuation of the output in the stop band depends on the order of polynomials of the transfer function of the circuit. For example a first order transfer function will produce an attenuation of $-20\,\mathrm{dB/decade}$ in the stop band whereas a second order transfer function will give $-40\,\mathrm{dB/decade}$ attenuation. Each additional order gives $-20\,\mathrm{dB/decade}$ steeper slope. Another well used

Fig. 3.27. Ideal and practical gain responses of (a) low-pass, (b) high-pass, (c) band-pass and (d) band-reject filters.

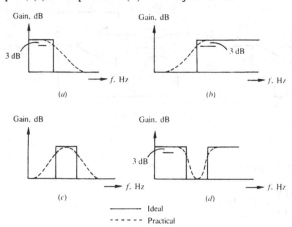

approximation gives the Chebyshev response in which the magnitude response in the pass band may have ripples ranging typically from 0.1 dB to 3 dB. Another commonly used approximation gives a maximally flat response but with a delay. The response is known as the Bessel response.

3.2.6.1 First-order circuits

Low-pass filters. In the first place we must select an appropriate transfer function for the filter synthesis and then design a circuit to yield that transfer function. Let us consider the gain response of a simple low-pass filter as shown in Fig. 3.28(a). The gain of the circuit, A_{vo} is constant from d.c. to some specified cutoff frequency f_0 and then at higher frequencies the gain falls off at a rate of 20 dB/decade. This characteristic can be realised by a first-order transfer function given by

$$A_v(s) = A_{vo}/(1 + s/\omega_0)$$

Fig. 3.28(a). The gain response of a first order low-pass filter.

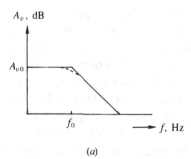

(a)

or,

$$A_v(j\omega) = A_{vo}/(1 + j\omega/\omega_0)$$

or,

$$A_v(jf) = A_{vo}/(1 + jf/f_0) \tag{3.38}$$

Now the magnitude of the above function can be written as follows,

$$G = |A_v(jf)| = \frac{A_{vo}}{\sqrt{\left(1 + \left(\dfrac{f}{f_0}\right)^2\right)}} \tag{3.39}$$

and its phase

$$\phi = \tan^{-1}0 - \tan^{-1}f/f_0 = -\tan^{-1}\frac{f}{f_0} \tag{3.40}$$

The transfer function given by (3.38) can be achieved by the circuit shown in Fig. 3.28(b), since the voltage gain of this circuit can be written as

$$A_v = -\frac{R_2}{R_1} \cdot \frac{1}{(1 + j2\pi fCR_2)} \tag{3.41}$$

which is of the same form as (3.38). Now comparing the two equations (3.38) and (3.41) we may write

$$A_{vo} = -R_2/R_1$$

or

$$|A_{vo}| = R_2/R_1 \tag{3.42}$$

and

$$f_0 = 1/2\pi CR_2 \tag{3.43}$$

Fig. 3.28(b). A first-order low-pass filter circuit.

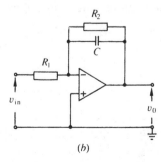

(b)

Fig. 3.29(a). The gain response of a first-order high-pass filter.

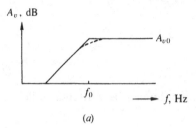

(a)

Therefore the circuit works as a low-pass filter whose cutoff frequency is determined by the circuit components in the feedback path of the operational amplifier.

High-pass filters. This type of filter as the name implies, attenuates low frequency and passes high frequency signals. The amplitude response of a first-order circuit is shown in Fig. 3.29(a). By observing the characteristic we can say that a zero in the transfer function of the gain curve occurs at zero frequency and the gain curve rises at a rate of 20 dB/decade up to the cutoff frequency f_0 where a pole occurs to make the gain constant (0 dB/decade) from that frequency onwards. The transfer function for this characteristic should therefore be of the form

$$A_v(s) = A_{vo}s/(\omega_0 + s) = A_{vo}/(1 + \omega_0/s)$$

or,

$$A_v(j\omega) = \frac{A_{vo}}{1 + \omega_0/j\omega} = \frac{A_{vo}}{1 - j\omega_0/\omega}$$

or,

$$A_v(jf) = \frac{A_{vo}}{1 - j(f_0/f)} \tag{3.44}$$

Now the magnitude of the above function can be written as follows,

$$G = |A_v(jf)| = \frac{A_{vo}}{\sqrt{\left(1 + \left(\frac{f_0}{f}\right)^2\right)}} \tag{3.45}$$

and its phase

$$\phi = \tan^{-1}0 - \tan^{-1}\left(-\frac{f_0}{f}\right) = -\tan^{-1} -\frac{f_0}{f} \tag{3.46}$$

Fig. 3.29(b). A first-order high-pass filter circuit.

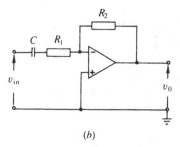

(b)

Now let us consider the circuit shown in Fig. 3.29(b). Its voltage gain is given by

$$A_v = -R_2/(R_1 + 1/j2\pi fC) = -\frac{R_2}{R_1\left(1 + \dfrac{1}{j2\pi fCR_1}\right)} \tag{3.47}$$

Equation (3.47) is of similar form to (3.44) and by comparing the two we may write

$$A_{vo} = -R_2/R_1$$

or

$$|A_{vo}| = R_2/R_1 \tag{3.48}$$

and

$$f_0 = 1/2\pi CR_1 \tag{3.49}$$

Therefore the circuit illustrated in Fig. 3.29(b) is a high-pass filter whose cutoff frequency is determined by the circuit elements in the input path of the operational amplifier.

Band-pass filters. A first-order band-pass filter can be obtained by cascading a high-pass first-order filter section with a low-pass first-order filter section. The cutoff frequency of the low-pass filter must be greater than that of the high-pass filter and the difference between the frequencies at 3 dB point will determine the bandwidth of the band-pass filter.

Band-reject filters. A first-order band-reject filter can be obtained in a similar fashion as above by cascading a low-pass filter section with a high-pass filter section, but in this case the cutoff frequency of the low-pass filter must be smaller than that of the high-pass filter and the difference between the frequencies at 3 dB point will determine the bandwidth of the band-reject filter.

Fig. 3.30. A general infinite gain multiple feedback second-order circuit.

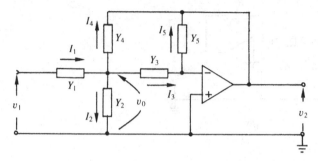

3.2.6.2 Infinite gain multiple feedback (IGMF) second-order circuits

Multiple feedback circuits have second-order voltage transfer functions. A general circuit that may be used to realise voltage transfer functions with a single pair of complex conjugate poles, and with zeros located only at the origin of the complex frequency plane or at infinity is shown in Fig. 3.30. In order to obtain the transfer function of the circuit we shall use Y parameters for the passive elements in the circuit. Each of the elements represents a single resistor or a single capacitor. By observing the circuit we may write

$$v_1 = \frac{1}{Y_1}I_1 + v_0 \tag{3.50}$$

Now let us assume that the input impedance and the differential mode gain of the operational amplifier are both infinite, then we may write the following expressions

$$I_1 = I_2 + I_3 + I_4 \tag{3.51}$$

$$I_2 = Y_2 v_0 \tag{3.52}$$

$$I_4 = Y_4(v_0 - v_2) \tag{3.53}$$

and

$$I_3 = Y_3 v_0 = -Y_5 v_2 = I_5 \tag{3.54}$$

From (3.50)–(3.54), the voltage transfer function of the multiple feedback circuit can be obtained

$$\frac{v_2}{v_1} = \frac{-Y_1 Y_3}{Y_5(Y_1 + Y_2 + Y_3 + Y_4) + Y_3 Y_4} \tag{3.55}$$

Fig. 3.31. A second-order low-pass filter circuit.

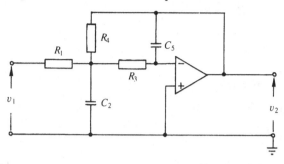

The passive elements of the circuit can be readily chosen so as to obtain voltage transfer functions for low-pass, high-pass or band-pass filters. *Low-pass filters.* The voltage transfer function that will give the required response is given by

$$\frac{v_2}{v_1}(S) = \frac{A_{vo}\omega_0^2}{S^2 + \omega_0\alpha S + \omega_0^2} \tag{3.56}$$

where A_{vo} is a positive real constant which specifies the gain in the pass band, i.e. the d.c. and low frequency gain in this case. The transfer function has two complex conjugate poles which will cause the gain to fall at a rate of 40 dB/decade in the stop band. Now to make the general circuit work as a low-pass filter we must choose the passive elements such that the voltage transfer function of (3.55) becomes similar in form to that shown in (3.56). It is obvious from these equations that the elements Y_1, Y_3 and Y_4 must be resistors whereas in order to generate an s^2 term in the denominator Y_2 and Y_5 must be capacitors. The circuit thus obtained is shown in Fig. 3.31. Substituting $Y_1 = 1/R_1$, $Y_2 = sC_2$, $Y_3 = 1/R_3$, $Y_4 = 1/R_4$ and $Y_5 = sC_s$ in (3.55) we obtain

$$\frac{v_2}{v_1}(s) = \frac{-\dfrac{1}{R_1 R_3}\cdot\dfrac{1}{C_2 C_5}}{s^2 + s\dfrac{1}{C_2}\left(\dfrac{1}{R_1} + \dfrac{1}{R_3} + \dfrac{1}{R_4}\right) + \dfrac{1}{R_3 R_4 C_2 C_5}} \tag{3.57}*$$

By comparing (3.57) with (3.56) we get

$$A_{vo}\omega_0^2 = \frac{1}{R_1 R_3 C_2 C_5} \tag{3.58}$$

* The negative sign in the equation means that the circuit produces a signal inversion.

$$\alpha\omega_0 = \frac{1}{C_2}\left(\frac{1}{R_1} + \frac{1}{R_3} + \frac{1}{R_4}\right) \tag{3.59}$$

and

$$\omega_0{}^2 = \frac{1}{R_3 R_4 C_2 C_5} \tag{3.60}$$

From (3.58) and (3.60)

$$A_{vo} = R_4/R_1 \tag{3.61}$$

The magnitude response of a second-order low-pass filter for several values of α (where $\alpha/2$ is the damping factor) is shown in Fig. 3.32. The gain of the filter is unity and it has been plotted in dB against the normalised frequency in a log scale. It can be noted that in the pass band the 'maximally flat' Butterworth response occurs when $\alpha = \sqrt{2}$.

From (3.56),

$$s = [-\omega_0\alpha \pm \sqrt{(\omega_0{}^2\alpha^2 - 4\omega_0{}^2)}]/2$$

Therefore we can say by observing the above expression that α should be less than 2 in order to yield complex conjugate poles.

Worked example 3.19
Design a second-order, low-pass filter having a maximum gain of 10 and a 3 dB frequency of 1 kHz. The filter should have a maximally flat response in the pass band and an input impedance of $10\,\mathrm{k\Omega}$.

3.32. The gain response of a second-order low-pass filter for several values of α where $\alpha/2$ is the damping factor.

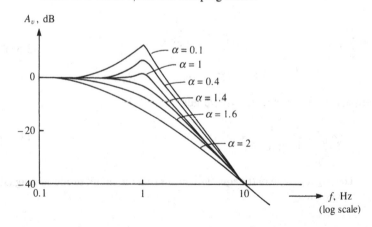

Solution

Consider the circuit of Fig. 3.31, $\alpha = \sqrt{2}$ for maximally flat response. For input impedance of $10\,k\Omega$, $R_1 = 10\,k\Omega$ (say) therefore from (3.61)

$$R_4 = 10 \times 10\,k\Omega = 100\,k\Omega$$

From (3.59), substituting for α, ω_0, R_1 and R_4 we obtain,

$$\sqrt{2} \times 2\pi \times 1000 = \frac{1}{C_2}\left(\frac{1}{10 \times 10^3} + \frac{1}{R_3} + \frac{1}{100 \times 10^3}\right)$$

Choose $C_2 = 0.1\,\mu F$ and the above expression becomes

$$\sqrt{2} \times 2\pi \times 1000 \times 0.1 \times 10^{-6} = 10^{-4} + \frac{1}{R_3} + 10^{-5}$$

$$\therefore R_3 = 1.285\,k\Omega$$

From (3.60):

$$C_5 = \frac{1}{R_3 R_4 C_2 \omega_0{}^2}$$

$$= \frac{1}{1.285 \times 10^3 \times 100 \times 10^3 \times 0.1 \times 10^{-6} \times 2\pi \times 2\pi \times 10^6}$$

$$= 1.973\,nF$$

If impractical values of R_3 and C_5 are obtained then we can revise the choice of C_2 and repeat the calculations.

Worked example 3.20

Design a fourth-order, low-pass, maximally flat response filter with a gain of 20 and corner frequency 4 kHz.

Solution

From the table given in Appendix E, the transfer function for fourth-order L.P.F. having corner frequency of 1 rad/s and gain of 1 is

$$\frac{1}{(s^2 + 0.765\,s + 1)(s^2 + 1.848\,s + 1)}$$

This transfer function can be achieved by cascading two second-order filters having transfer functions

$$\frac{1}{s^2 + 0.765s + 1} \text{ and } \frac{1}{s^2 + 1.848s + 1}$$

First consider

$$\frac{1}{s^2 + 0.765s + 1}$$

From (3.61)

$$R_4/R_1 = 1$$

Choose

$$R_4 = 10\,\text{k}\Omega$$

$$\therefore R_1 = 10\,\text{k}\Omega$$

From (3.59):

$$\omega_0 \alpha = \frac{1}{C_2}\left(\frac{1}{R_1} + \frac{1}{R_3} + \frac{1}{R_4}\right)$$

Choose

$$C_2 = 0.1 \times 10^{-2}\,\text{F}$$

$$\therefore 0.765 = \frac{1}{0.1 \times 10^{-2}}\left(10^{-4} + \frac{1}{R_3} + 10^{-4}\right)$$

or

$$R_3 = 1.7\,\text{k}\Omega$$

From (3.60):

$$C_5 = \frac{1}{1.7 \times 10^3 \times 10 \times 10^3 \times 0.1 \times 10^{-2}} = 59\,\mu\text{F}$$

The frequency level is changed to $\omega_0 = 2\pi \times 4 \times 10^3$ by dividing the capacitor values by the new ω_0.
Thus

$$C_2 = 0.1 \times 10^{-2} \div 2\pi \times 4 \times 10^3 = 0.039\,\mu\text{F}$$

and

$$C_5 = 59 \times 10^{-6} \div 2\pi \times 4 \times 10^3 = 2.3\,\text{nF}$$

$\therefore$ The first stage is as shown in Fig. ex.3.20(a).
Now consider

$$\frac{1}{s^2 + 1.848s + 1}$$

Fig. ex.3.20(a)

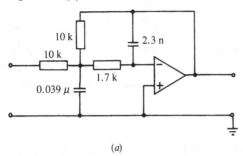

(a)

From (3.61):

$$R_4 = R_1 = 10\,k\Omega \text{ (assuming } R_4 = 10\,k\Omega)$$

From (3.59):

$$1.848 = \frac{1}{0.1 \times 10^{-2}}\left(10^{-4} + \frac{1}{R_3} + 10^{-4}\right)$$
$$\text{(assuming } C_2 = 0.1 \times 10^{-2})$$

or

$$R_3 = 607\,\Omega$$

From (3.60):

$$C_5 = \frac{1}{607 \times 10 \times 10^3 \times 0.1 \times 10^{-2}} = 0.165 \times 10^{-3}\,F$$

For $\omega_0 = 2\pi \times 4 \times 10^3\,\text{rad s}^{-1}$ C_2 becomes 0.039 μF and C_5 becomes 6.5 nF. Therefore the second stage is as shown in Fig. ex.3.20(b).

A gain of 20 is obtained by using an amplifier (here an inverting amplifier has been used).

Fig. ex.3.20(b)

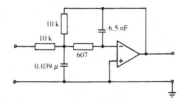

(b)

Fig. ex.3.20(c)

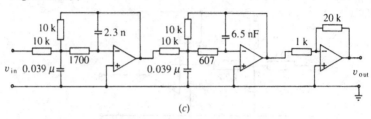

(c)

Figure ex.3.20(c) shows the complete low-pass filter circuit.
High-pass filters. The voltage transfer function that will give the required response is given by

$$\frac{v_2}{v_1}(s) = \frac{-A_{vo}s^2}{s^2 + \alpha\omega_0 s + \omega_0^{\,2}}$$ (3.62)

where A_{vo} is a positive real constant which specifies the gain in the pass band, i.e. the high frequency gain in this case. The transfer function has two zeros at the origin and two complex conjugate poles the vector of which gives the cutoff frequency.

Now we must choose the passive elements of the general circuit in such a way that the voltage transfer function of (3.55) becomes similar in form to that shown in (3.62). Again it is obvious from the two equations that the elements Y_1 and Y_3 must be capacitors in order to generate an s^2 term in the numerator. After a close examination of the two equations we may also conclude that the s^2 term in the denominator can only be generated by having a capacitor for Y_4. The other two passive elements must be resistors. The high-pass filter circuit thus obtained is shown in Fig. 3.33(a). Substituting $Y_1 = sC_1$, $Y_2 = 1/R_2$, $Y_3 = sC_3$, $Y_4 = sC_4$ and $Y_5 = 1/R_5$ in (3.55) gives

$$\frac{v_2}{v_1}(s) = \frac{-s^2 C_1/C_4}{s^2 + s(C_1 + C_3 + C_4)\dfrac{1}{C_3C_4R_5} + \dfrac{1}{C_3C_4R_2R_5}}$$ (3.63)

Now comparing (3.62) with (3.63) we may write the following expressions

$$A_{vo} = C_1/C_4$$ (3.64)

$$\alpha\omega_0 = (C_1 + C_3 + C_4)/C_3C_4R_5$$ (3.65)

and

$$\omega_0 = \sqrt{\left(\frac{1}{C_3C_4R_2R_5}\right)}$$ (3.66)

Fig. 3.33(a). A second-order high-pass filter circuit and (b) its gain response.

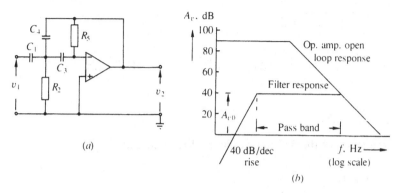

(a)

(b)

The gain response of the high-pass filter circuit is shown in Fig. 3.33(b). It should be noted that at very high frequency (much above f_H) the closed loop gain curve meets the open loop gain curve of the operational amplifier and follows the latter.

Worked example 3.21

Design a second-order, maximally flat, high-pass filter with a gain of 6 in the pass band and a corner frequency of 8 kHz. A distorted 2 kHz sinusoidal input has a third harmonic content of 1.2 V_{PTP}, what would be its amplitude at the output of the filter?

Solution

Consider the circuit of Fig. 3.33(a)

For 'maximally flat' response in the pass band $\alpha = \sqrt{2}$. Choose $C_1 = C_3 = 0.01\ \mu F$.

From (3.64)

$$C_4 = \frac{C_1}{A_{vo}} = \frac{0.01 \times 10^{-6}}{6} = 1.7\ nF$$

From (3.65)

$$R_5 = \frac{C_1 + C_3 + C_4}{C_3 C_4 \alpha \omega_0}$$

$$= \frac{(0.01 + 0.01 + 0.0017) \times 10^{-6}}{0.01 \times 10^{-6} \times 0.0017 \times 10^{-6} \times \sqrt{2} \times 2\pi \times 8 \times 10^3} = 17.96\ k\Omega$$

From (3.66)

$$R_2 = \frac{1}{R_5 C_3 C_4 \omega_0^2}$$

$$= \frac{1}{17.96 \times 10^3 \times 0.01 \times 10^{-6} \times 0.0017 \times 10^{-6} \times (2\pi \times 8 \times 10^3)^2}$$

$$= 1.3\,k\Omega$$

The frequency of the 3rd harmonic is 6 kHz and gain

$$G = \left| \frac{A_{vo}}{1 - (\omega_0/\omega)^2 - j\alpha\dfrac{\omega_0}{\omega}} \right|$$

$$= \frac{A_{vo}}{\sqrt{\{[1-(\omega_0/\omega)^2]^2 + \alpha^2(\omega_0^2/\omega^2)\}}} = \frac{6}{\sqrt{\{[1-(\tfrac{8}{6})^2]^2 + 2 \times (\tfrac{8}{6})^2\}}}$$

$$= 2.95$$

Therefore 3rd harmonic output $= 1.2 \times 2.95 = 3.54\,V_{PTP}$

Band-pass filters

In the voltage transfer function of this filter we must have a pair of complex conjugate poles the vector of which is equal to the centre frequency of the band-pass filter and a zero located at the origin. Therefore the following expression will give the required response.

$$\frac{v_2}{v_1}(s) = \frac{-A_{vo}\omega_0 s}{s^2 + \alpha\omega_0 s + \omega_0^2} \tag{3.67}$$

where A_{vo} is a positive real constant and $A_{vo}/\alpha = M_o$ is the magnitude of the gain in the pass band. The magnitude response rises at a rate of 20 dB/decade from very low frequency to the centre frequency, f_0 after which, due to a pair of poles at f_0, the gain drops at the rate of -20 dB/decade. In effect a -40 dB/decade slope has been added to the $+20$ dB/decade slope at f_0. Now if we choose $Y_1 = 1/R_1$, $Y_2 = 1/R_2$, $Y_3 = sC_3$, $Y_4 = sC_4$ and $Y_5 = 1/R_5$ for the basic circuit and then substitute these in (3.55) then the voltage transfer function becomes

$$\frac{v_2}{v_1}(s) = \frac{-s/C_4 R_1}{s^2 + s\dfrac{1}{R_5}\left(\dfrac{1}{C_3} + \dfrac{1}{C_4}\right) + \dfrac{1}{R_5 C_3 C_4}\left(\dfrac{1}{R_1} + \dfrac{1}{R_2}\right)} \tag{3.68}$$

Fig. 3.34. A second-order band-pass filter circuit and its gain response.

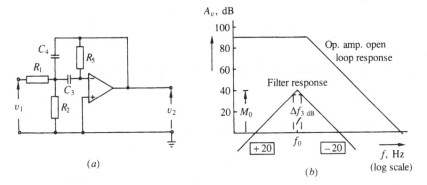

(a)

(b)

which is similar to (3.67). Figure 3.34 shows the band pass filter circuit and its magnitude response. By comparing (3.67) with (3.68) we get

$$\omega_0 = \sqrt{\left[\frac{1}{R_5 C_3 C_4} \left(\frac{1}{R_1} + \frac{1}{R_2} \right) \right]} \qquad (3.69)$$

$$A_{vo}\omega_0 = \frac{1}{C_4 R_1} \qquad (3.70)$$

$$\alpha\omega_0 = \frac{1}{R_5} \left(\frac{1}{C_3} + \frac{1}{C_4} \right) \qquad (3.71)$$

and

$$\alpha = 1/Q = \sqrt{\left[\frac{1}{R_5 \left(\frac{1}{R_1} + \frac{1}{R_2} \right)} \right]} \cdot \left[\sqrt{\left(\frac{C_3}{C_4} \right)} + \sqrt{\left(\frac{C_4}{C_3} \right)} \right] \quad (3.72)$$

where Q is the quality factor. For band-pass filters it is an important parameter and is given by the ratio of the centre frequency, ω_0 to the 3 dB bandwidth of the filter, $\Delta\omega_{3\,dB}$. Filters with high Q (i.e. $Q \gg 1$, typical value of Q is 10, it can be as high as 100) are referred to as narrow-band filters.

In practice the resistor R_1 is very much larger than the resistor R_2 and we can use R_2 to trim the Q factor. Then the centre frequency can be adjusted by simultaneously varying R_2 and R_5 by the same percentage.

Worked example 3.22
Design a second-order infinite gain multiple feedback band-pass filter with a centre frequency of 20 kHz and a Q of 5. The gain should be 2 at the centre frequency.

Solution

Given

$$M_0 = 2, \quad \therefore A_{vo} = \alpha M_0 = \frac{M_0}{Q} = \frac{2}{5}$$

Consider the circuit of Fig. 3.34(a). Choose $C_3 = C_4 = 0.001\ \mu\text{F}$.
From (3.70)

$$R_1 = \frac{1}{A_{vo}\omega_0 C_4} = \frac{1}{\frac{2}{5} \times 2\pi \times 20 \times 10^3 \times 0.001 \times 10^{-6}} = 19.9\ \text{k}\Omega$$

From (3.71)

$$R_5 = \frac{1}{\alpha\omega_0}\left(\frac{1}{C_3} + \frac{1}{C_4}\right) = \frac{1}{\frac{1}{5} \times 2\pi \times 20 \times 10^3}\left(\frac{10^6}{0.001} + \frac{10^6}{0.001}\right)$$

$$= 79.6\ \text{k}\Omega$$

From (3.69)

$$R_2 = 1/(R_5 C_3 C_4 \omega_0^2 - 1/R_1)$$

$$= 1\Big/\Big[\,79.6 \times 10^3 \times 0.001 \times 10^{-6} \times 0.001 \times 10^{-6}$$

$$\times (2\pi \times 20 \times 10^3)^2 - \frac{1}{19.9 \times 10^3}\Big]$$

$$= 829\ \Omega.$$

Sallen and Key circuits. Another configuration that can be used to construct
low-pass, high-pass and band-pass filters is the Sallen and Key (VCVS –
voltage controlled voltage source) circuit. Figure 3.35 shows a general

Fig. 3.35. A general noninverting configuration (Sallen and Key) that
can be used to construct second-order low-pass and high-pass filters.

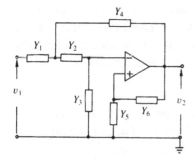

Parameters Filters	Y_1	Y_2	Y_3	Y_4	Y_5	Y_6
Low pass	R_1	R_2	C_3	C_4	R_5	R_6
High pass	C_1	C_2	R_3	R_4	R_5	R_6

Fig. 3.36. A biquad low-pass filter circuit.

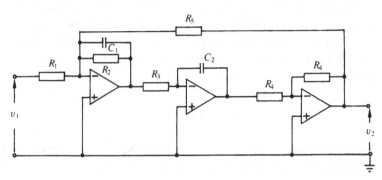

noninverting configuration that can be used to construct second-order low-pass and high-pass filters of both the Butterworth and Chebychev responses. However, the IGMF (infinite gain multiple feedback) design is more popular because it requires one less component than the VCVS design, and it has low output impedance.

Multiple operational amplifier biquad circuits. A low-pass filter of the popular biquad design is shown in Fig. 3.36. Although this design requires three operational amplifiers, its gain and cutoff frequency can be easily adjusted by varying the resistances R_1 and R_3 respectively. From the figure it can be noticed that both inverted and noninverted outputs are available from this circuit. The gain of the filter is given by the ratio of R_5 to R_1. The gain is changed by varying R_1, whereas, both the gain and the cutoff frequency are varied by adjusting R_5.

3.2.6.3 Switched-capacitor filters

Switched-capacitor (SC) filters are analogue data sampling systems comprising high quality MOS capacitors, operational amplifiers and switches. A capacitor and two switches simulate the circuit behaviour of a resistor as shown in Fig. 3.37. The capacitor is switched between terminals

Fig. 3.37. A switched-capacitor circuit and the MOSFET gate pulses.

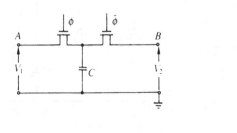

A and *B* by the two switches. When the capacitor is connected to terminal *A* it charges to the input voltage V_1, then when it is connected to the other terminal *B* the capacitor discharges. The amount of charge which flows into or from terminal *B* is therefore equal to $C(V_2 - V_1)$. If the switches are operated at a clock frequency f_c, then the average current flow from *A* to *B* will be

$$\frac{C(V_2 - V_1)}{T_c} = C(V_2 - V_1)f_c \qquad (3.73)$$

where T_c is the switching period. By observing (3.73) we may say that the circuit is equivalent to a 'resistor' having a value

$$R = \frac{1}{Cf_c} \qquad (3.74)$$

This equivalence is only valid when the input signal frequency is much lower than the clock frequency. If they are of the same order, then sampled data techniques have to be used and the input signal should be band-limited according to the sampling theorem. A switched-capacitor resistor of 1 MΩ value can be implemented by switching a 1 pF capacitor at a rate of 1 MHz. This capacitor will need a silicon area of approximately 0.01 mm², whereas, to implement this resistor using a polysilicon line or diffusion, an area 100 times larger will be required. Another advantage of using SC techniques is that the elimination of resistors reduces power consumption. Figure 3.38(*a*) shows a simple first order *RC* low-pass passive filter. The transfer function of the circuit is given by

$$\frac{V_2}{V_1}(S) = \frac{\dfrac{1}{SC_1}}{R_1 + 1/SC_1} = \frac{1}{1 + SC_1 R_1} \qquad (3.75)$$

Fig. 3.38. (*a*) A simple first-order *RC* low-pass passive filter and (*b*) its switched-capacitor implementation.

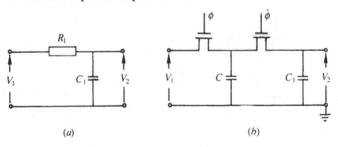

(*a*)   (*b*)

Fig. 3.39. (*a*) A simple first-order *RC* high-pass passive filter and (*b*) its switched-capacitor implementation.

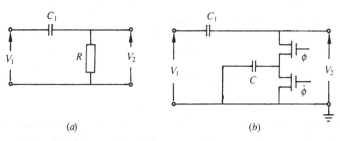

(*a*)   (*b*)

and the 3-dB bandwidth of this filter is

$$f_{3\,\mathrm{dB(RC)}} = \frac{1}{2\pi C_1 R_1} \tag{3.76}$$

The switched-capacitor implementation of this circuit is shown in Fig. 3.38(b). By substituting the effective resistance of the switched-capacitor *C* from (3.74) into (3.76) we obtain the 3-dB bandwidth of the switched-capacitor filter as

$$f_{3\,\mathrm{dB(SC)}} \simeq \frac{1}{2\pi} \cdot f_C\left(\frac{C}{C_1}\right) \tag{3.77}$$

It can be observed from (3.77) that the 3-dB bandwidth can be varied simply by changing the clock frequency.

A simple first order *RC* high pass passive filter is shown in Fig. 3.39(*a*) and the switched-capacitor implementation of this circuit is shown in Fig. 3.39(*b*).

Switched-capacitor low-pass filters. A second-order switched-capacitor low-pass filter will now be developed starting from a passive *LC* network. Fig. 3.40 shows a second-order low-pass passive filter network. Its transfer function is

$$\frac{V_{\mathrm{out}}}{V_{\mathrm{in}}}(S) = \frac{\dfrac{1}{SC}}{R + SL + \dfrac{1}{SC}} = \frac{1}{S^2 LC + SCR + 1} \tag{3.78}*$$

---

*   The values of the inductor and the capacitor for this passive prototype filter can be found from standard design tables for a normalized low pass filter.

Fig. 3.40. A second-order low-pass passive filter circuit.

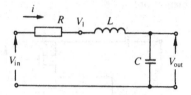

We can describe the network by equations containing integrations as shown in (3.79). Switched-capacitor integrators which will be discussed in the next section can then be used to realise the network. By observing the network in Fig. 3.40 we may write

$$\left.\begin{aligned} V_1 &= V_{\text{in}} - Ri \\[2mm] i &= \frac{1}{SL}(V_1 - V_{\text{out}}) \\[2mm] \text{and} \quad V_{\text{out}} &= \frac{1}{SC}i \end{aligned}\right\} \tag{3.79}$$

Since the operational amplifiers which will be used to realize this circuit are voltage controlled voltage sources we multiply the current i of the expressions (3.79) by a scaling resistance R_x to convert it into voltage, $V = iR_x$. Thus the expressions (3.79) become

$$\left.\begin{aligned} V_1 &= V_{\text{in}} - \frac{R}{R_x}V \\[2mm] V &= \frac{R_x}{SL}(V_1 - V_{\text{out}}) \\[2mm] \text{and} \quad V_{\text{out}} &= \frac{1}{SR_xC}V \end{aligned}\right\} \tag{3.80}$$

These equations can be represented by the schematic arrangement shown in Fig. 3.41 which has two integrators. The transfer functions of the integrators are

$$\frac{V}{V_1 - V_{\text{out}}}(S) = \frac{1}{S\dfrac{L}{R_x}} \tag{3.81}$$

and

$$\frac{V_{\text{out}}}{V}(S) = \frac{1}{SR_xC} \tag{3.82}$$

Fig. 3.41. The schematic diagram of low-pass filter circuit using active integrators.

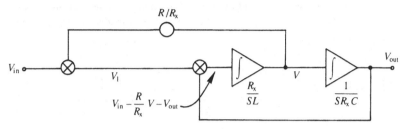

And they have unity-gain bandwidths R_x/L and $1/R_xC$ respectively in radians/s.

Switched-capacitor integrators. The basic switched-capacitor concept can be used to replace the resistor of an integrator circuit as shown in Fig. 3.42(a). The MOS switches turn on and off with the rise and fall of the clock pulses. Let us assume that a step voltage is applied to the input terminal. During the period $T_c/2$ when capacitor C is connected to the input terminal it charges toward the input voltage and the output voltage remains constant. During the next half cycle of the clock when capacitor C is connected to the inverting input terminal of the operational amplifier the charge transfers from C to C_1 immediately and the voltage V_{out} changes accordingly. The output voltage is given by

$$V_{out} = -\frac{Cf_c}{SC_1}V_{in} \qquad (3.83)$$

and has the waveform shown in Fig. 3.42(b), the time constant being C_1/Cf_c.

Fig. 3.42. (a) A switched-capacitor integrator circuit and (b) its output voltage for a step input.

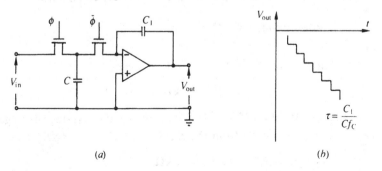

(a)   (b)

Fig. 3.43. A switched-capacitor implementation of the schematic diagram of Fig. 3.41. Here differential integrators have been used, the first one acts as a subtracting integrator whereas the second one acts as a single input integrator since one input is grounded.

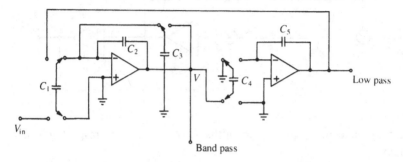

A switched capacitor implementation of the schematic arrangement of Fig. 3.41 is shown in Fig. 3.43. This circuit simultaneously provides second-order low-pass and band-pass filtering, from different points in the circuit.

The RS Components RSMF10 is a dual switched capacitor filter integrated circuit fabricated in CMOS technology. In each channel it has two non-inverting switched capacitor integrators and other necessary components to enable one to construct a complete second order state variable type active filter capable of providing low-pass, band-pass, and high-pass outputs simultaneously.

Worked example 3.23

Design a fourth-order low-pass filter with a gain of 45 dB and the cutoff frequency at 1.5 kHz using a RS MF10 dual switched-capacitor filter integrated circuit. (See Appendix for schematic diagram of RS MF10 and the design equations.)

Solution

Let us choose a two-stage filter each stage having the same cutoff frequency. From the information sheet of RS MF10 circuit,

$$\text{Cutoff frequency} = \frac{f_{CL}}{100}$$

or

clock frequency = 100 × cutoff frequency

$$= 100 \times 1.5\,\text{kHz} = 150\,\text{kHz}$$

Now the gain 45 dB = 177.8 = 12 × 14.7 (say), i.e. the first stage has a gain of 12 therefore if $R_1 = 1\,\text{k}\Omega$ from the expression $H_{OLP} = -R_2/R_1$ we obtain

$$R_2 = |H_{OLP}|R_1 = 12 \times 1\,\text{k} = 12\,\text{k}\Omega$$

Fig. ex.3.23

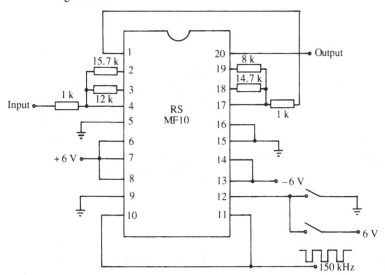

For maximally flat response assume $Q = \dfrac{1}{0.765}$ for the first stage, then

$$R_3 = QR_2 = \frac{1}{0.765} \times 12\,\text{k} = 15.7\,\text{k}\Omega$$

The second stage has a gain of 14.7.

For this stage we may again choose $R_1 = 1\,\text{k}\Omega$, then

$$R_2 = |H_{\text{OLP}}|R_1 = 14.7 \times 1\,\text{k} = 14.7\,\text{k}\Omega$$

and assuming $Q = \dfrac{1}{1.848}$ for the second stage

$$R_3 = QR_2 = \frac{1}{1.848} \times 14.7\,\text{k} = 8\,\text{k}\Omega$$

It should be noted that the first stage gives underdamped response whereas the second stage gives overdamped response, thus overall frequency response being the Butterworth one.

Figure ex.3.23 shows how the above components have to be connected to the RS MF10 IC.

3.2.7   Pulse and waveform generators

3.2.7.1 Astable multivibrators

These circuits, also known as free running multivibrators, can be used to produce square waves. It should be mentioned here that there are

Fig. 3.44. An astable multivibrator circuit.

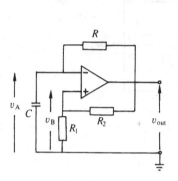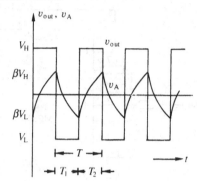

digital ICs which are cheaper and much better in performance. An astable multivibrator switches repetitively between two unstable states. In the circuit of Fig. 3.44 the operational amplifier together with the positive feedback resistors R_1 and R_2 form a basic comparator circuit. The output of the circuit saturates at V_H and V_L. From the circuit it can be observed that the output will switch states from positive to negative and vice versa when the voltage at the inverting input terminal, v_A becomes equal to βV_H and βV_L respectively where β is the feedback factor and given by

$$\beta = \frac{R_1}{R_1 + R_2} \tag{3.84}$$

Let us consider that the output voltage is at positive saturation V_H at $t=0$. The capacitor C charges towards V_H through R. When the potential at the inverting input terminal, v_A reaches the potential of the noninverting input terminal, $v_B = \beta V_H$, the positive feedback forces regenerative switching. The output changes from positive saturation to negative saturation V_L. The voltage across the capacitor falls exponentially towards V_L as shown in the diagram and when the potential v_A reaches the potential, $v_B = \beta V_L$, the output changes from negative saturation to positive saturation. The process is repeated and a square wave is obtained at the output.

We can find the frequency of oscillations from the 'charging up' and 'charging down' times of the capacitor C. The time periods T_1 and T_2 are given by

$$T_1 = CR \ln \frac{V_L - \beta V_H}{V_L - \beta V_L} \tag{3.85}$$

and

$$T_2 = CR \ln \frac{V_H - \beta V_L}{V_H - \beta V_H} \tag{3.86}$$

respectively.

It is apparent from the expressions that if the positive saturation voltage is equal in magnitude to that of the negative saturation voltage,

$$T_1 = T_2 = CR \ln \frac{1+\beta}{1-\beta} = CR \ln \left(1 + \frac{2R_1}{R_2} \right) \tag{3.87}$$

Thus the frequency of oscillations is

$$f = \frac{1}{T_1 + T_2} = \frac{1}{2CR \ln \left(1 + \frac{2R_1}{R_2} \right)} \tag{3.88}$$

In the above circuit we have a 50% duty cycle, i.e. the mark–space ratio is unity. By having different magnitudes of positive and negative supply voltages we may have mark–space ratios other than unity. However, the circuit shown in Fig. 3.45 is more commonly used where two diodes are connected between the output of the circuit and the capacitor via the potential divider, $R_M + R_S$. Here when the output is positive the diode D_1 is forward biased and diode D_2 is reverse biased. So the capacitor charges up via the lower part of the potential divider with resistance R_M until the circuit switches over, and the output is at negative saturation. The diode D_2 will now be forward biased and diode D_1 will be reverse biased, hence the capacitor will discharge through the upper portion of the potential divider

Fig. 3.45. An astable multivibrator circuit with variable mark-space ratio and its output waveform.

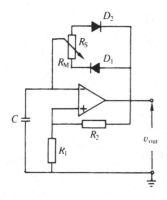
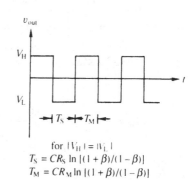

for $|V_H| = |V_L|$

$T_S = CR_S \ln |(1 + \beta)/(1 - \beta)|$

$T_M = CR_M \ln |(1 + \beta)/(1 - \beta)|$

with resistance R_S. By adjusting the potential divider the mark–space ratio of the output waveform can be varied. It should be noted that since the total resistance $R_M + R_S$ remains constant at all conditions, the frequency of oscillations will not change.

Worked example 3.24

Design a circuit which will provide a $10\,\text{kHz}$ square wave signal of ± 10 volts magnitude with a duty cycle of 40%.

Solution

Consider the circuit of Fig. 3.45

$$T_M : T_S = 4:6 \text{ or, } T_M = 0.67 T_S$$

also

$$T_M + T_S = 1/(10 \times 10^3 \text{s})$$

From the above two expressions

$$T_S = \frac{1}{1.67 \times 10 \times 10^3}$$

$$= 0.06 \times 10^{-3}\,\text{s}$$

Choose

$$C = 0.01\,\mu\text{F}, \quad R_1 = 1\,\text{k}\Omega, \quad R_2 = 10\,\text{k}\Omega$$

Then from (3.84)

$$\beta = \frac{R_1}{R_1 + R_2} = \frac{1}{11} = 0.09$$

From Fig. 3.45

$$R_S = \frac{T_S}{C \ln \dfrac{1+\beta}{1-\beta}} = \frac{0.06 \times 10^{-3}}{0.01 \times 10^{-6} \ln \dfrac{1.09}{0.91}}$$

$$= 33.2\,\text{k}\Omega$$

and

$$R_M = \frac{T_M}{C \ln \dfrac{1+\beta}{1-\beta}} = \frac{0.06 \times 10^{-3} \times 0.67}{0.01 \times 10^{-6} \ln \dfrac{1.09}{0.91}}$$

$$= 22.3\,\text{k}\Omega$$

Several zener diodes with total zener voltage of 9.4 V with forward voltage of 0.6 V connected across the output terminal and ground back to back will give ± 10 V amplitude. A resistor has to be inserted between the output terminal of the operational amplifier and the zener diodes to limit the current flowing through the diodes.

3.2.7.2 Monostable or one shot multivibrators

The monostable multivibrator circuit can produce a rectangular pulse of predetermined duration when a trigger pulse is applied to the input terminal. This circuit as shown in Fig. 3.46 is similar in form to that of the astable multivibrator with the exception of a diode connected here in parallel with the capacitor in such a way that the inverting input terminal remains always positive at a potential equal to the forward voltage drop of the diode. The network consisting of C_T, R_T and D_2 provides a negative trigger pulse to the noninverting input terminal whatever the duration of the input pulse.

Let us consider that the amplifier output is held at positive saturation voltage, V_H. The potential divider R_1 and R_2 feeds back βV_H to the noninverting terminal. The inverting terminal is clamped at approximately 0.6 V (forward voltage drop of the diode) by the diode D_1. Thus the potential at the inverting input terminal is less than that at the noninverting input terminal, and the output is held at V_H. Now suppose a negative pulse, v_T, of a magnitude greater than βV_L is applied at X and the potential becomes negative at the noninverting input terminal. Thus v_A becomes greater than v_B and the output switches to the negative saturation voltage V_L. Now R_1 and R_2 feed back a negative voltage to the noninverting terminal. The capacitor starts discharging from $+0.6$ V towards V_L. However, when the potential v_A reaches the value βV_L the inverting input

Fig. 3.46. A monostable multivibrator circuit and the trigger and output waveforms.

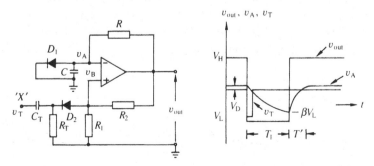

terminal becomes more negative than the noninverting input terminal and the output switches back to V_H.

The time period T_1 is given by,

$$T_1 = CR \ln \frac{V_L - V_D}{V_L - \beta V_L} \tag{3.89}$$

where

$$\beta = \frac{R_1}{R_1 + R_2}$$

We may neglect the forward voltage drop of the diode, V_D and the time period of the rectangular pulse is

$$T_1 = CR \ln \left(1 + \frac{R_1}{R_2} \right) \tag{3.90}$$

As soon as the output changes state to the positive value, the capacitor starts charging towards V_H. The time period T' is the time the capacitor takes to charge to V_D. During this period a negative trigger pulse may not initiate the circuit. Therefore in order to obtain a train of rectangular pulses from a train of trigger pulses the time interval between two pulses should be at least equal to the sum of the recovery time T' and the time period T_1. The recovery time T' is given by

$$T' = CR \ln \frac{V_H - \beta V_L}{V_H - V_D} \tag{3.91}$$

A differentiating circuit formed by R_T and C_T blocks any d.c. voltage or slowly varying signal and ensures that only rapidly varying pulses can initiate the monostable circuit.

Worked example 3.25

Several trigger pulses are consecutively applied to the circuit of Fig. 3.46. What should be the minimum time interval between any two of these pulses so that the circuit functions properly? Given that $C = 0.1\ \mu F$, $R = 470\ k\Omega$, $R_1 = 1\ k\Omega$, $R_2 = 10\ k\Omega$ and the forward voltage drop of the diodes is 0.6 volt. The output saturation voltages are ± 12 volts.

Solution
From (3.89)

$$T = CR \ln \frac{V_L - V_D}{V_L - \beta V_L} = 0.1 \times 10^{-6} \times 470 \times 10^3 \ln \frac{-12 - 0.6}{-12 + 0.09 \times 12}$$

$$= 6.72\ ms$$

Fig. 3.47. (a) A triangular waveform generator made of a comparator and an integrator and (b) their output voltage characteristic.

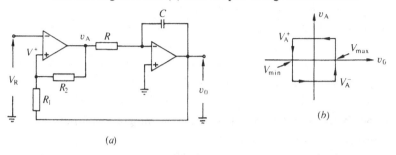

(a)

(b)

and from (3.91),

$$T' = CR \ln \frac{V_H - \beta V_L}{V_H - V_D} = 0.1 \times 10^{-6} \times 470 \times 10^3 \ln \frac{12 - 0.09(-12)}{12 - 0.6}$$

$$= 6.46 \, \text{ms}$$

$\therefore$ time interval $= T + T' = 13.18 \, \text{ms}$.

3.2.7.3 Triangular wave generator

The circuit shown in Fig. 3.47 generates a triangular wave with the help of a regenerative comparator and an integrator. Let us assume that at a particular instant the output of the comparator is at negative saturation; since the operational amplifier is connected in the inverting mode the output of the integrator will rise linearly until the feedback voltage to the noninverting terminal of the comparator becomes equal to V_R. The positive feedback immediately forces regenerative switching and the output of the comparator goes to positive saturation. At this stage the output voltage of the integrator v_0 reaches V_{max} (say) and starts decreasing. It will fall linearly until at a value of $v_0 = V_{min}$, the feedback voltage to the comparator input terminal is equal to V_R.

Using the superposition theorem, from the figure we may write

$$V_R = V^+ = \frac{V_A^- \cdot R_1}{R_1 + R_2} + \frac{V_{max} R_2}{R_1 + R_2}$$

$$\therefore \quad V_{max} = \frac{R_1 + R_2}{R_2} V_R - \frac{R_1}{R_2} V_A^- \tag{3.92}$$

Also

$$V_R = V^+ = \frac{V_A^+ R_1}{R_1 + R_2} + \frac{V_{min} R_2}{R_1 + R_2}$$

$$\therefore \quad V_{min} = \frac{R_1 + R_2}{R_2} V_R - \frac{R_1}{R_2} V_A^+ \tag{3.93}$$

It can be seen from (3.92) and (3.93) that if $V_R = 0$, then

$$V_{max} = -\frac{R_1}{R_2}V_A{}^-$$

and

$$V_{min} = -\frac{R_1}{R_2}V_A{}^+$$

When the comparator output reaches $V_A{}^+$ the integrator output starts to fall from V_{max} at a rate $-V_A/CR$ volts/s.

The time taken by the integrator to reach the voltage V_{min} from V_{max} is

$$T = \frac{CR}{V_A}(V_{max} - V_{min}) = \frac{CR}{V_A}\left(-\frac{R_1}{R_2}V_A{}^- + \frac{R_1}{R_2}V_A{}^+\right)$$

$$= \frac{CR}{V_A} \cdot 2 \cdot \frac{R_1}{R_2} \cdot V_A = 2C\frac{RR_1}{R_2}$$

Therefore the frequency of oscillation is

$$1/2T = R_2/4CRR_1 \tag{3.94}$$

Worked example 3.26

Draw the output waveform of the circuit of Fig. 3.47 if $V_R = 0$, $R_1 = 3.9\,k\Omega$, $R_2 = 10\,k\Omega$, $R = 68\,k\Omega$, $C = 0.01\,\mu F$ and $V_A{}^+ = -V_A{}^- = 12\,V$.

Solution

$$V_{max} = -\frac{R_1}{R_2} \cdot V_A{}^- = -\frac{3.9 \times 10^3}{10 \times 10^3} \times -12 = 4.68\,V$$

$$V_{min} = -\frac{R_1}{R_2} \cdot V_A{}^+ = -\frac{3.9 \times 10^3}{10 \times 10^3} \times 12 = -4.68\,V$$

From (3.94) the frequency,

$$f = 1/2T = \frac{R_2}{4CRR_1}$$

$$= \frac{10 \times 10^3}{4 \times 0.01 \times 10^{-6} \times 68 \times 10^3 \times 3.9 \times 10^3}$$

$$= 943\,Hz$$

and

$$T = 0.53\,ms$$

Fig. ex. 3.26

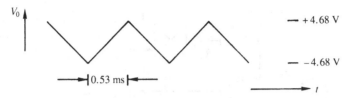

3.2.8   Inverse function generators

It has been shown in the earlier section that two logarithmic amplifiers in conjunction with one antilogarithmic amplifier can multiply two quantities expressed as voltages or currents. Various inverse function generators can be built by using such or any other multipliers in conjunction with operational amplifiers. Let us first consider the basic configuration of an inverse function generator which is shown in Fig. 3.48. The rectangle in the feedback path of the circuit showing $f(\ )$ represents a function generator whose inverse may be obtained at the output of the circuit. Assuming ideal operational amplifier performance we may write

$$I_1 + I_2 = 0$$

$$I_2 = \frac{f(v_{out})}{R}$$

and

$$I_1 = v_{in}/R$$

From the above three equations

$$f(v_{out}) = -v_{in}$$

or,

$$v_{out} = f^{-1}(-v_{in}) \tag{3.95}$$

Fig. 3.48. The basic configuration of an inverse function generator.

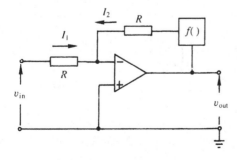

Fig. 3.49. A divider circuit.

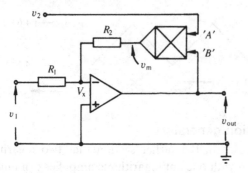

Thus the inverse of the function $f(\ )$ is obtained by using this circuit. It will be shown now that by having a multiplier in the feedback path various circuits which can perform division, square rooting and RMS operations can be obtained.

3.2.8.1 Divider

In this circuit as shown in Fig. 3.49 the output is fed back to one of the two input terminals of the multiplier and one of the two input signals is applied to the other terminal of the multiplier. Thus the output of the multiplier itself is

$$v_m = A v_2 v_{out}$$

where A is the multiplier constant.

Now applying the Kirchhoff's laws to the input terminals of the operational amplifier we may write

$$\frac{v_1}{R_1} + \frac{v_m}{R_2} = 0$$

or

$$v_1/R_1 + \frac{A v_2}{R_2} v_{out} = 0$$

or

$$v_{out} = \frac{R_2}{A R_1} \cdot \frac{1}{v_2} (-v_1) = -\frac{1}{B} \left(\frac{v_1}{v_2} \right)$$

where B is $\dfrac{A R_1}{R_2}$. Therefore

$$v_{out} \propto (v_1 \div v_2) \tag{3.96}$$

Fig. 3.50. A square root circuit.

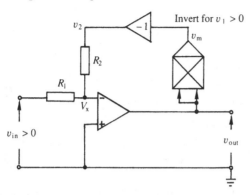

Thus a dividing circuit is obtained by using a multiplier in the feedback path. If the signal v_2 is negative, then positive feedback will take place and the circuit may be unstable. So the input signal v_2 must be positive, if not then an inverter should be used after the multiplier to make the v_m negative.

3.2.8.2 Square root circuit

In this circuit the output is fed back to the two input terminals of the multiplier simultaneously as shown in Fig. 3.50. The multiplier output is proportional to the square of the output voltage v_0, i.e., $v_m = A v_{out}^2$. Unless the polarity of v_m is changed (assuming the input signal, v_{in} is positive) a positive feedback will occur and the circuit would be unstable. Therefore an inverter must be placed after the multiplier. Now from the Fig. 3.50 we may write making usual assumptions for an ideal operational amplifier performance

$$v_2 = -A v_{out}^2 \tag{3.97}$$

and

$$\frac{v_{in}}{R_1} + \frac{v_2}{R_2} = 0 \tag{3.98}$$

From (3.97) and (3.98)

$$v_{out} = -\sqrt{\left(\frac{1}{A}\frac{R_2}{R_1} \cdot v_{in}\right)} = -k\sqrt{v_{in}} \tag{3.99}$$

For this circuit to be stable v_{out} should be negative and consequently the input signal has to be positive.

If the input signal is negative then the inverter following the multiplier is not needed. In this case

$$v_2 = A v_{\text{out}}^2 \tag{3.100}$$

and

$$-\frac{v_{\text{in}}}{R_1} + \frac{v_2}{R_2} = 0 \tag{3.101}$$

From (3.100) and (3.101) we get

$$v_{\text{out}} = -\sqrt{\left(\frac{1}{A} \cdot \frac{R_2}{R_1} v_{\text{in}}\right)} = -k\sqrt{v_{\text{in}}} \tag{3.102}$$

This circuit yields the same expression as in the case of a positive input signal. Thus we obtain the square root of a voltage by using a multiplier as a squarer in the feedback path.

3.2.8.3 Root mean square circuit

The operational amplifier acts as a summer to the outputs of three multipliers M_1, M_2 and M_3 which have multiplier constants A_1, A_2 and A_3 respectively. The input signals v_1 and v_2 are applied to multipliers M_1 and M_2, whereas, the output v_{out} is inputted to multiplier M_3 as shown in Fig. 3.51. From the figure we can easily write, after making usual assumptions,

$$v_3 = A_1 v_1^2$$

$$v_4 = A_2 v_2^2$$

$$v_6 = -A_3 v_{\text{out}}^2$$

Fig. 3.51. A root mean square circuit.

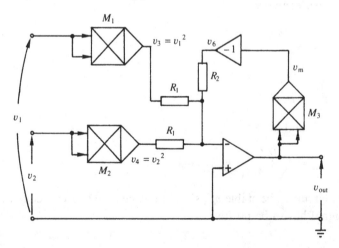

and

$$\frac{v_3}{R_1} + \frac{v_4}{R_1} + \frac{v_6}{R_2} = 0 \tag{3.103}$$

From the above equations we obtain, assuming $A_1 = A_2 = A_3$,

$$v_{\text{out}}^2 = \frac{R_2}{R_1}(v_1^2 + v_2^2)$$

or

$$v_{\text{out}} = -\sqrt{(R_2/R_1)} \cdot \sqrt{(v_1^2 + v_2^2)} = -k\sqrt{(v_1^2 + v_2^2)}$$

The voltages at terminals '3' and '4' are positive, and v_m is also positive, whether the input signals are positive or not. Thus v_{out} will always be negative.

Worked example 3.27

Show how operational amplifiers can be used in conjunction with multipliers to build a circuit which will provide the following function.

$$v_{\text{out}} \propto 3/2 \sqrt{(v_1^2 + v_2)}$$

Solution

A low cost integrated circuit analogue multiplier type RS 1495 can be used for dividing, squaring and square rooting.

Fig. ex.3.27

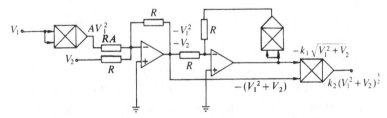

Summary

1. The full bandwidth of an operational amplifier circuit can be calculated from the formula

$$f = \frac{S}{2\pi V_{\text{op}}}$$

where S is the slew rate of the operational amplifier and defined as the maximum change in the output voltage per unit of time and V_{op} is the peak-to-peak output voltage.

2. Error voltages appear at the output of an operational amplifier circuit due to the presence of input offset voltages and input bias currents. The worst case error voltage at the output of an inverting amplifier due to the input offset voltages and bias currents can be calculated using the formula

$$V_{os(total)} = V_{io}\frac{R_1 + R_2}{R_1} + I_b{}^- R_2$$

In an inverting amplifier the input bias currents can be compensated by adding a resistor at the noninverting input terminal. The worst case error voltage at the output then becomes

$$V_{os(total)} = V_{io}\frac{R_1 + R_2}{R_1} + I_{os}R_2$$

where I_{os} is the difference between $I_b{}^+$ and $I_b{}^-$ and typically is one fifth of the bias currents.

3. A circuit can be compensated for both the input offset voltage and the bias current, but only at one temperature. Any change in temperature will cause an error in the output voltage. The error is given by

$$E = \left(\frac{\Delta V_{io}}{\Delta T}\right) \cdot \Delta T \cdot \frac{R_1 + R_2}{R_1} + \left(\frac{\Delta I_{os}}{\Delta T}\right) \cdot \Delta T \cdot R_2$$

4. The voltage gain of an ideal operational amplifier should be independent of frequency, but in reality it decreases at a rate of 20 dB/decade as the frequency increases due to the stray and junction capacitances present in the circuit.

5. An operational amplifier circuit is unstable, i.e. produces oscillations, if the loop gain of the circuit has a phase shift of 360° at unity gain. One can determine whether a circuit is stable or not by using a Bode plot of the operational amplifier's open loop gain curve and the closed loop gain curve and examining the rate of closure at their crossing. A stable circuit has a rate of closure of 20 dB/decade on both sides of the crossing (here the loop gain is unity since $\log A\beta = \log A - \log 1/\beta = 0$ dB), whereas, a rate of closure of 40 dB/decade on both sides of the crossing gives rise to an unstable circuit. A circuit with a minimum allowable phase margin of 45° has a maximum possible bandwidth and the rates of closure of 20 dB/decade on one side and 40 dB/decade on the other side of the crossing.

6. An unstable circuit can be made conditionally stable by adding a capacitor (lag compensation method) or a capacitor in series with a resistor

(limited lag compensation method) to the operational amplifiers. The limited lag compensation method gives larger bandwidth than the lag compensation method.

7. Many operational amplifiers are internally compensated by the manufacturers and can be used for any closed loop gain. However, their useful bandwidth is very narrow.

8. The output of many transducers is a very small differential voltage imposed on a large unwanted common-mode voltage which may be a d.c. level or a noise pick-up. It is possible to amplify the differential voltage only to an acceptable level with the help of an instrumentation amplifier which is basically a precision differential amplifier with a very high input impedance and gain.

9. An instrumentation amplifier may be balanced before its use to produce zero output voltage for zero input voltage, various changes produce an offset error voltage at the output. The error voltage is determined using the following expression.

$$E_{os(total)} = \pm \frac{\Delta E_{os}}{\Delta T} \cdot \Delta T \pm \frac{\Delta E_{os}}{\Delta V_s} \cdot \Delta V_s \pm \frac{\Delta E_{os}}{\Delta t} \cdot \Delta t$$

$$\pm R_s \frac{\Delta I_{os}}{\Delta T} \cdot \Delta T \pm \Delta E_n$$

where $\dfrac{\Delta E_{os}}{\Delta T}, \dfrac{\Delta E_{os}}{\Delta V_s}, \dfrac{\Delta E_{os}}{\Delta t}$ are offset voltage drifts with temperature, power supply variation and time,

$\dfrac{\Delta I_{os}}{\Delta T}$ is the offset current drift with temperature,

ΔE_n is the output noise voltage

and

R_s is the source resistance.

10. Comparators are used to compare the magnitude of a signal with that of a reference voltage which may be fixed or varying. Operational amplifiers without any feedback act as comparators because of their very large gain. Regenerative comparators which change states very rapidly but at different input voltage levels (hysteresis) have many applications one of which is to clean noisy digital signals.

11. Diodes or transistors are used in the feedback path of the operational amplifier circuits to construct logarithmic amplifiers. For antilogarithmic amplifiers they are placed in the input path. These are themselves used to

build circuits to provide at the output various functions of one or more input signals.

12. Another use of operational amplifiers is in analogue computers which can be used to solve linear differential equations describing various physical and electrical systems.

13. One of the most important applications of operational amplifiers is in active filters which do not require any inductors. Low-pass, high-pass, band-pass and band-reject filters mainly with Butterworth response have been discussed in this text. The rate of decrease in gain in the stop band depends on the order of the polynomials of the voltage transfer functions of the filter circuits. Switched-capacitor filters have also been discussed, these are available in integrated circuit form and the cutoff or centre frequencies can be varied easily by varying the frequency of switching the capacitors which act as resistors in the circuits.

15. Various waveform and function generator circuits employing operational amplifiers have been briefly discussed.

Problems

3.1. A 741 type operational amplifier has a slew rate of $0.5\,V/\mu s$ and is used in an inverting amplifier with a gain of 250. Calculate the time required for the output to reach 50% of its steady state value if a step voltage of $10\,mV$ is inputted to the amplifier. If the input is a sinusoidal signal of $100\,kHz$, what is the maximum amplitude of the output voltage without distortion?

3.2. A step input of $60\,mV$ is fed to an inverting amplifier which has a gain of 80. If the time required for the output to reach within 0.05% of its final value is $8\,\mu s$, calculate the slew rate of the operational amplifier used in the circuit.

3.3. A sinusoidal signal is applied to the circuit mentioned in Problem 3.2. The typical slew rate of the 741 type operational amplifier is $0.5\,V/\mu s$. Calculate the frequency of the signal if the maximum amplitude of the input without having any distortion of the output is $20\,mV$.

3.4. The worst case offset voltage at the output of the operational amplifier in the circuit of Fig. 3.2(c) is $\pm 0.4\,V$. If $R_1 = 2.2\,k\Omega$ and $R_2 = 56\,k\Omega$, what will be the maximum input offset voltage?

3.5. In Fig. 3.2(c), the maximum input offset voltage is $2.5\,mV$. If the circuit has a gain of unity, calculate the maximum output offset voltage. If the operational amplifier has $I_b = 60\,nA$ and $I_{os} = 15\,nA$ and the resistors R_1 and R_2 are $1\,k\Omega$ and $47\,k\Omega$ respectively, estimate the total maximum output offset voltage without and with bias current compensation. What will be the value of the compensating resistor?

3.6. The worst case values of the output offset voltage of the operational amplifier in the circuit of Fig. 3.3(a), with and without bias current compensation are 54 mV and 58 mV respectively. If the compensating resistor is 2.117 kΩ, $R_1 = 2.2$ kΩ and the input offset voltage is 2 mV, calculate the input offset current and the bias current of the operational amplifier.

3.7. The error voltage at the output of the circuit of Fig. 3.3(a) due to a temperature change of 20 °C from room temperature is 8 mV. Calculate the input offset voltage temperature coefficient at room temperature given that $R_1 = 2.2$ kΩ, $R_2 = 100$ kΩ and $\Delta I_{os}/\Delta T = 1$ nA/°C at room temperature.

3.8. The bridge circuit shown in Fig. 3.11 has a differential voltage $\Delta V = 10$ mV for $V = 4.5$ V and $R = 10$ kΩ. Calculate ΔR. The output of the bridge is connected to an instrumentation amplifier as shown in Fig. 3.12. If $n = 25$, $R_B = 20$ kΩ and $R_A = 1$ kΩ, what is the output voltage of the amplifier?

3.9. An integrated circuit instrumentation amplifier has the following specifications at a gain of 800: $\Delta E_{os}/\Delta T = \pm 8$ mV/°C, $\Delta I_{os}/\Delta T = \pm 2$ nA/°C, $\Delta E_{os}/\Delta t = \pm 4$ mV/24 h, $\Delta E_n = 5$ mV rms and $\Delta E_{os}/\Delta V_s = \pm 50$ mV/V. If this amplifier is balanced to zero output for zero input before being connected to the bridge described in Worked example 3.9 and the power suply needed for the amplifier is ± 12 V, after what period of time, the offset error voltage would be ± 250 mV? Assume that the temperature change during this period of time is 20 °C and the variation in the power supply is 2%.

3.10. An instrumentation amplifier connected to a bridge circuit with $R = 8.2$ kΩ and $\Delta R = 800$ Ω, has the following specifications at a gain of 1500: $\Delta E_{os}/\Delta T = \pm 10$ mV/°C, $\Delta E_{os}/\Delta V_s = \pm 50$ mV/V, $\Delta E_{os}/\Delta t = \pm 57.9$ nV/s, $\Delta I_{os}/\Delta T = \pm 1$ nA/°C and $\Delta E_n = 6$ mV rms. What would be the output offset error voltage of the amplifier after 4 weeks operation if the temperature fluctuation during this period is ± 10 °C and the ± 15 V power supply varies by $\pm 5\%$?

3.11. The low frequency differential mode gain of the operational amplifier used in the circuit of Fig. 3.3(a) is 100 dB and has corner frequencies at 10 kHz and 100 kHz. What is the minimum gain the amplifier could have if the circuit is lag compensated with the help of a capacitor of 1.592 μF and the output resistance of the preceding stage at the compensating point is 1 kΩ?

Fig. P.3.13

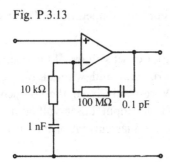

3.12. The limited lag frequency compensation is applied to the circuit of Problem 3.11 to make it critically stable. Find the values of the compensating elements. What is the advantage of using this method of compensation over the lag compensation method?

3.13. The operational amplifier used in the circuit shown in Fig. P.3.13 has corner frequencies at 3 kHz and 30 kHz, and a low frequency differential mode gain of 120 dB. Examine the stability of the circuit and if necessary apply limited lag frequency compensation. The output resistance of the preceding stage at the compensating point is 15 kΩ.

3.14. What are the maximum and minimum input voltages required to switch over the output voltage levels of the circuit of Fig. 3.15(a)? Given that $V_H = 15$ V, $V_L = -15$ V, $R_1 = 1$ kΩ, $R_2 = 6.8$ kΩ and the reference voltage, $V_{ref} = 4$ V.

3.15. A sinusoidal signal of ± 10 V amplitude and frequency 1 kHz is applied to the input of the circuit of Fig. 3.15(a). Draw the output voltage waveform if the operational amplifier has a slew rate of 1 V/μs. Given: $R_1 = 1$ kΩ, $R_2 = 15$ kΩ, $V_{ref} = 3$ V, $V_H = 18$ V and $V_L = -18$ V.

3.16. An operational amplifier has a slew rate of 2 V/μs and is connected as a comparator circuit as shown in Fig. 3.15(a) with an input of ± 5 V, 10 kHz triangular wave. Draw the output voltage waveform.

3.17. Draw the output voltage waveform for a 100 Hz square wave input of $4V_{PTP}$ amplitude applied to the circuit of Fig. 3.17(b). The silicon diodes have leakage currents of 100 nA and $R_1 = 1$ kΩ and $R_2 = 10$ kΩ.

3.18. Design a full wave rectifier to deliver 12 V d.c. with a 0.4 V_{PTP} ripple into a resistive load of 2 kΩ, the input signal being the same as that given in Worked example 3.13.

3.19. Two silicon diodes having leakage currents of 8 μA are connected in a circuit as shown in Fig. 3.17(b). If R_1/R_2 has a ratio of 0.21, draw the output voltage waveform for a 150 Hz sinusoidal input voltage of 0.707 V rms.

Design a full wave rectifier to deliver 10 V d.c. with a ripple of 0.2 V_{PTP} fed into a resistive load of 3 kΩ.

3.20. What should be the input voltage of the logarithmic amplifier circuit of Fig. 3.21 to give an output voltage of -300 mV. Given $R = 10$ kΩ and $I_S = 2$ pA at an ambient temperature of 27 °C.

3.21. The logarithmic amplifier shown in Fig. 3.21 has $R = 100$ kΩ and $I_S = 1$ pA at an ambient temperature of 19 °C. Calculate the output voltage for an input voltage of 100 mV. If a voltage amplifier with a gain of 100 is connected in between the input signal and the logarithmic amplifier, what would be the output voltage of the logarithmic amplifier?

3.22. What is the value of resistor R_5 in the circuit of Fig. 3.22, if the output voltage is -300 mV for an input voltage of 100 mV at an ambient temperature of 27 °C? Given: $R_6 = 33$ kΩ.

3.23. Design an analogue computation circuit to solve the differential equation

$$3\frac{d^3x}{dt^3} - 8\frac{d^2x}{dt^2} + 5\frac{dx}{dt} + 6x + 4 = 0$$

3.24. The fuel consumption of a motor vehicle with a fuel tank having a volume of 1.5 m³ and a height of 0.3 m is at a rate of 0.03 m³/s. Design an analogue computer to indicate the level of fuel in the tank.

3.25. The low pass filter shown in Fig. 3.31 has a maximally flat response in the pass band with $\alpha = \sqrt{2}$. Calculate the maximum gain, the 3-dB frequency and the value of C_5. Given: $R_1 = 12$ kΩ, $R_3 = 1.2$ kΩ, $R_4 = 150$ kΩ and $C_2 = 0.22$ μF.

3.26. The fourth order, low pass, maximally flat response filter shown in Fig. P.3.26 has a gain of unity in the pass band. Find the cutoff frequency of the filter and the values of capacitors C_5 and C_5'.

Fig. P.3.26

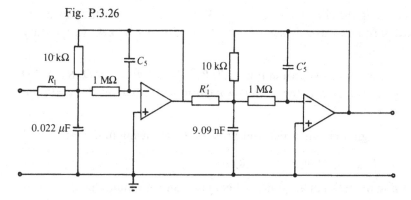

3.27. Design a second-order low pass filter having an input impedance of 5 kΩ, a maximum gain of 10 and a 3-dB frequency at 1 kHz. It should have a maximally flat response in the pass band.

3.28. A distorted 3 kHz sinusoidal input having a second harmonic content of $1.6 V_{PTP}$ is applied to a second-order, maximally flat, high-pass filter circuit as shown in Fig. 3.33(a). The filter has a corner frequency of 12 kHz. If the amplitude of the third harmonic content at the output of the filter is $3.2V_{PTP}$, find the value of C_3, C_4 and R_2. Assume $C_1 = 0.022 \mu F$ and $R_5 = 27 k\Omega$.

3.29 Design a second-order, maximally flat, high-pass filter with a gain of 10 in the pass band and a corner frequency of 4 kHz. A distorted 1 kHz sinusoidal voltage which has a $0.2V_{PTP}$ third harmonic component is inputted to the filter. What would be the amplitude of the third harmonic component at the output of the filter?

3.30 A second-order infinite gain multiple feedback band pass filter circuit as shown in Fig. 3.34(a) has a centre frequency of 10 kHz. Find the gain at the centre frequency and the value of Q, if $R_2 = 980 \Omega$, $R_5 = 270 k\Omega$ and $C_3 = C_4 = 0.001 \mu F$.

3.31. Design a second-order infinite gain multiple feedback band-pass filter with a centre frequency of 100 kHz and a Q of 5. The gain at the centre frequency is 1.25 dB. Use the circuit shown in Fig. 3.34(a).

3.32. Design a fourth-order, low-pass, switched capacitor filter using a RS MF10 type integrated circuit. The filter should have a gain of 40 dB and the cutoff frequency at 1 kHz. See the appendix for the data of the integrated circuit.

3.33. The square wave signal generator shown in Fig. 3.45 has $R_1 = 2.2 k\Omega$, $R_2 = 12 k\Omega$, $C = 0.022 \mu F$, $R_M = 20 k\Omega$ and $R_S = 33 k\Omega$. Calculate the output frequency.

3.34. Using the circuit shown in Fig. 3.45, design a generator which will produce square-wave signals of ± 10 V magnitude at 15 kHz with a duty cycle of 30%.

3.35. In the circuit shown in Fig. 3.47, $R_1 = 3.9 k\Omega$, $R_2 = 22 k\Omega$, $R = 51 k\Omega$ and $C = 10 nF$. Draw the output waveform of this circuit if $V_R = 0$ and $V_A{}^+ = -V_A{}^- = 12$ V.

3.36. Design a circuit which will provide the following function:

$$V_{out} = k_1 \sqrt{(av_1 + bv_2{}^2)} - k_2 v_3{}^2$$

by using operational amplifiers in conjunction with multipliers.

4

Oscillators

Objectives

At the end of the study of this chapter a student should be:

1. familiar with the conditions of sinusoidal oscillations
2. able to understand the principles of LC oscillators
3. familiar with the principle of operations of RC oscillators
4. able to construct crystal oscillators.

In electronics periodic signals of various shapes such as sinusoidal, triangular, rectangular and pulses are often needed to perform different types of operation. Although the term oscillator is generally referred to a generator of sinusoidal signals, while a rectangular-wave generator is more commonly known as a multivibrator, an oscillator generates an a.c. output signal without requiring any form of input signal. Sinusoidal oscillators are mainly used in radio frequency transmitters and receivers. Square-wave or pulse generators are used in almost every type of digital equipment. In this chapter we shall mainly concentrate on sinusoidal oscillators.

4.1 Criteria for oscillation

An oscillator circuit is basically an amplifier, but a positive feedback from its output to the input enables it to sustain the output without the need for an external signal. Fig. 4.1 illustrates an amplifier with a gain of A and a feedback network having a feedback fraction of β.

Fig. 4.1. An amplifier with a gain of A and a feedback network having a feedback fraction of β.

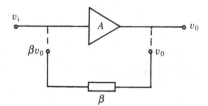

Firstly let us consider the amplifier on its own. Suppose an input signal v_i is applied to it, its output v_o will be then given by,

$$v_o = Av_i \qquad (4.1)$$

Now if we connect the feedback network to the amplifier as shown in the figure and can somehow make the input v_i equal to the output of the feedback network so that they have identical magnitude, phase and frequency, then the circuit will not need any external signal. In other words, the circuit will act as an oscillator if

$$v_i = \beta v_o \qquad (4.2)$$

Substituting for v_o from (4.1) into (4.2) we obtain

$$v_i = Av_i\beta$$
$$A\beta = 1 \qquad (4.3)$$

The above relationship is known as the Barkhausen criterion.

Thus the magnitude of the loop gain should be unity and the phase shift of the loop gain should be $0°$ or a multiple of $360°$ at the frequency of interest, since both the magnitude and phase shift of the loop gain are functions of frequency.

If we substitute the value of the loop gain into the gain expression for a closed loop circuit applying positive feedback

$$A_f = A/(1 - A\beta) \qquad (4.4)$$

where A_f is the gain of the feedback circuit, then

$$A_f = A/(1 - 1) = \infty$$

A gain of infinity implies that there is an output voltage even when no input signal is present, i.e., the circuit oscillates.

Worked example 4.1

In the circuit of Fig. 4.1 the gain of the forward amplifier A is frequency dependent and given by $A = -9 \times 10^6/j\omega$. If the feedback fraction β is $6 \times 10^3/(3 \times 10^3 + j\omega)^2$ will the circuit oscillate? If so, at what frequency oscillations occur?

Solution

The loop gain

$$A\beta = \frac{-9 \times 10^6 \times 6 \times 10^3}{j\omega(3 \times 10^3 + j\omega)(3 \times 10^3 + j\omega)}$$

$$= \frac{-54 \times 10^9}{j\omega \times 9 \times 10^6 - 6 \times \omega^2 \times 10^3 - j\omega^3} \tag{1}$$

So that the phase shift of the loop gain is 0, the imaginary part of the denominator should be zero.

$\therefore$ $j(\omega \times 9 \times 10^6 - \omega^3) = 0$

or

$\omega = 3 \times 10^3 \, \text{rad/s}$

Substituting this in (1) we have

$$A\beta = \frac{-54 \times 10^9}{-6 \times 10^3 \times 9 \times 10^6} = 1$$

The conditions for oscillations are satisfied. The circuit will oscillate at 477.7 Hz.

4.2   *LC* oscillator

The most common oscillators are of the tuned radio frequency type with inductance L and capacitance C. They produce a sinusoidal output at the resonant frequency of the tuned LC circuit. If the tuned circuit is placed in the feedback path of a feedback circuit, then the latter satisfies the Barkhausen criterion at the resonant frequency of the tuned circuit. Therefore, oscillations only occur at one frequency.

Figure 4.2 shows a general form of LC oscillator circuits. The active device may be a bipolar transistor, a field effect transistor or an operational amplifier. Let us assume that the device has an infinite input impedance, a gain A_v and an output resistance R_o. The feedback network comprises a

Fig. 4.2. A general form of LC oscillator circuit.

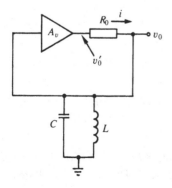

capacitor C and an inductor L. From the diagram we may write, assuming that there is no output current,

$$v_o' = v_o + R_o i$$

$$= v_o + R_o(v_o/j\omega L + v_o j\omega C)$$

$$= v_o(1 + R_o j\omega C + R_o/j\omega L) \tag{4.5}$$

The loop gain of the feedback circuit is given by

$$A\beta = A_v \cdot v_o/v_o' \tag{4.6}$$

From (4.5) and (4.6) we may write

$$A\beta = A_v/(1 + R_o j\omega C + R_o/j\omega L)$$

$$= A_v/(1 + jR_o(\omega C - 1/\omega L)) \tag{4.7}$$

In order to satisfy the phase shift condition for oscillation the imaginary quantity in the denominator of (4.7) should be equal to zero. Equating R_o the output resistance of the forward amplifier to zero in the imaginary part makes the feedback network gain independent of frequency, and the circuit oscillates at any frequency, therefore the quantity inside the bracket of the imaginary part,

$$\omega C - 1/\omega L = 0$$

or

$$f = 1/2\pi\sqrt{(LC)} \tag{4.8}$$

Thus oscillation only takes place when the feedback circuit is at resonance.

Worked example 4.2

What value of L is needed with a C of 47 pF for a tuned radio frequency oscillator frequency of 22.7 MHz?

Solution
From (4.8)

$$f_0 = 1/2\pi\sqrt{(LC)}$$

or

$$L = 1/4\pi^2 f_o^2 C = 1/4\pi^2 (22.7 \times 10^6)^2 \times 47 \times 10^{-12} = 1.04\,\mu\text{H}$$

The Hartley oscillator

The Hartley oscillator has a tuned LC circuit with a tapped coil for inductive feedback as shown in Fig. 4.3. The circuit oscillates at the

Fig. 4.3. The Hartley oscillator.

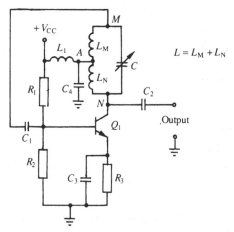

$$L = L_M + L_N$$

frequency $1/2\pi\sqrt{(LC)}$, C is a variable capacitor and can be adjusted to produce oscillations of different frequencies.

Transistor Q_1 forms a common emitter amplifier with the resistors R_1, R_2 and R_3 determining its d.c. bias. The tuned circuit forms the collector load and C_3 is the emitter bypass capacitor. The d.c. voltage is supplied to the collector of the transistor via the tap at point A on the tuning inductor $L_M + L_N$. The radio frequency choke L_1 does not allow the oscillator frequency to reach the d.c. power supply. Capacitor C_4 bypasses the radio frequency to ground. Therefore, for the oscillator signal the point A is effectively at ground potential. Thus the voltage V_{MA} is opposite in polarity to the voltage V_{NA} which is in fact the output of the circuit. Since there is a phase shift of 180° between the base and the collector of the transistor, the total phase shift of the complete loop is 360° or 0°. Therefore the circuit is an oscillator that produces an a.c. output at the resonant frequency of the tuned circuit. The feedback fraction of the circuit is given by V_{MA}/V_{NA} which is usually of the order of 0.3.

The frequency range for this type of oscillator is from 100 kHz to a few MHz. The restriction in the range is mainly due to the physical size of the tuning capacitor and the tapped coil. At low frequency, a tuning capacitor may be too large, whereas, in the very high frequency range it may be too difficult to tap a very small coil.

The Colpitts oscillator

The principle of operation of this type of oscillator is similar to that of the Hartley oscillator except that in this case capacitive feedback is used instead of inductive feedback. So the mark of the Colpitts oscillator is a

Fig. 4.4. The Colpitts oscillator.

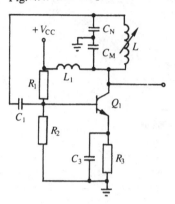

capacitive voltage divider as shown in Fig. 4.4. The capacitors C_M and C_N connected across the inductor L form the resonant circuit. Since the junction of C_M and C_N is grounded, V_{CN}, the voltage developed across C_N is exactly opposite in phase to V_{CM}, the voltage developed across C_M. V_{CN} is fed back to the base of the transistor via the coupling capacitor C_1. L_1 is a radio frequency choke providing a low-impedance path for the d.c. supply but acting as an open circuit to the oscillator output.

The Colpitts oscillator is tuned usually by varying the value of L since the capacitance of the resonant circuit is divided. This can be achieved by either adjusting a variable inductor or switching in coils of different values. The Colpitts oscillator is used for low radio frequencies and also for the very high frequency band up to 300 MHz. Thus it covers the frequency ranges which the Hartley oscillator can not for the reasons mentioned earlier.

Worked example 4.3

The transistor in the Colpitts oscillator of Fig. 4.4 has a mutual conductance of 5×10^{-3} Siemens, a base-emitter resistance of 1.5 kΩ, a base spreading resistance of 100 Ω and an output impedance equal to 10 kΩ. Calculate the values for C_M and C_N which will just cause the circuit to oscillate at $10/2\pi$ MHz. Given $R_B = R_1 /\!/ R_2 = 1.3$ kΩ and $L = 1$ mH.

Solution

The high frequency small signal equivalent circuit of the oscillator is as shown in Fig. ex.4.3(*a*):
At radio frequency the RF choke has very high impedance, simplifying the circuit of Fig. ex.4.3(*a*) to give Fig. ex.4.3(*b*)

where

$$R = R_B /\!/ (r_{bb}' + r_{be}')$$

Fig. ex.4.3

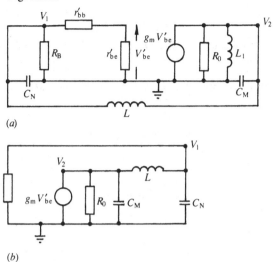

(a)

(b)

The collector impedance

$$Z_C = R_o \left/\left/ \frac{1}{j\omega C_M} \right.\right/\!\!\left/ \left[j\omega L + \frac{R}{1+j\omega C_N R} \right]\right.$$

$$= R_o \left/\left/ \frac{1}{j\omega C_M} \right.\right/\!\!\left/ \left(\frac{j\omega L - \omega^2 L C_N R + R}{1+j\omega C_N R} \right)\right.$$

$$= R_o \left/\left/ \frac{j\omega L - \omega^2 L C_N R + R}{1+j\omega C_N R - \omega^2 L C_M - j\omega^3 L C_N C_M R + j\omega R C_M} \right.$$

$$= \frac{R_o(R - \omega^2 L C_N R + j\omega L)}{R_o - R_o\omega^2 L C_M + R - \omega^2 L C_N R + j\omega R C_N R_o - j\omega^3 L C_N C_M R R_o + j\omega R C_M R_o + j\omega L}$$

Now, feedback fraction β

$$= \cfrac{\cfrac{\cfrac{1}{j\omega C_N} \times R}{R + \cfrac{1}{j\omega C_N}}}{j\omega L + \cfrac{\cfrac{1}{j\omega C_N} R}{R + \cfrac{1}{j\omega C_N}}} = \frac{R}{R + j\omega L - \omega^2 C_N L R}$$

Thus loop gain

$$A\beta = \frac{g_m R_o R}{R_o - R_o \omega^2 LC_M + R - \omega^2 LC_N R + j\omega C_N RR_o - j\omega^3 LC_N C_M RR_o + j\omega RC_M R_o + j\omega L}$$

For a phase shift of $0°$ the imaginary part in the denominator should be zero.

$\therefore$ $\quad\quad \omega[RR_o(C_M + C_N) + L - \omega^2 LC_N C_M RR_o] = 0$

or

$$\omega^2 = \frac{C_M + C_N}{LC_N C_M} + \frac{1}{C_N C_M RR_o} \simeq \frac{C_M + C_N}{LC_N C_M}$$

$\therefore$ $\quad\quad \omega = \sqrt{\left(\frac{C_M + C_N}{LC_N C_M}\right)}$

$$f = \frac{1}{2\pi}\sqrt{\left(\frac{C_M + C_N}{LC_N C_M}\right)} \quad\quad\quad\quad (1)$$

Also

$$A\beta = \frac{g_m R_o R}{R_o - R_o \omega^2 LC_M + R - \omega^2 LC_N R} = 1$$

or

$$g_m R_o R = R_o \frac{C_M}{C_N} + R \frac{C_N}{C_M}$$

or

$$5 \times 10^{-3} \times 10 \times 10^3 \times \frac{1.3 \times 10^3 \times 1.6 \times 10^3}{2.9 \times 10^3}$$

$$= 10 \times 10^3 \frac{C_M}{C_N} + \frac{1.3 \times 1.6 \times 10^6 C_N}{2.9 \times 10^3 C_M}$$

or

$$\frac{C_M{}^2}{C_N{}^2} - 3.59\frac{C_M}{C_N} + 0.072 = 0$$

$\therefore$ $\quad\quad \frac{C_M}{C_N} = \frac{3.59 \pm \sqrt{(12.88 - 0.288)}}{2} = 3.57 \text{ or } 0.02$

Substituting the values of f and L into (1)

$$\frac{10 \times 10^6}{2\pi} = \frac{1}{2\pi} \sqrt{\left/ \left(\frac{C_M + C_N}{1 \times 10^{-3} C_M C_N}\right)\right.}$$

or

$$\frac{C_M + C_N}{C_N C_M} = 10^{11}$$

Let us take

$$\frac{C_M}{C_N} = 3.57$$

then $C_N = 12.8\,\text{pF}$ and $C_M = 45.7\,\text{pF}$.

4.3   *RC* phase-shift oscillator

In order to obtain oscillation at audio frequencies from *LC* oscillators the component values have to be very large, which may be rather inconvenient for most applications. On the other hand *RC* oscillators use resistors and capacitors of acceptable sizes to produce the required phase shift at very low frequencies. Figure 4.5(*a*) shows a simple circuit which satisfies the conditions for oscillation. The transistor Q_1 together with R_C, R_E, R_3, R_4 and C_E form the forward amplifier. The voltage divider formed by R_3 and R_4 supplies the forward bias to the base of the transistor Q_1. The oscillator output from the collector of the transistor is fed back to the base of the transistor via the *RC* network. The forward amplifier causes a 180°

Fig. 4.5. (*a*) The *RC* phase-shift oscillator and

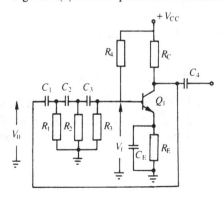

(*a*)

Fig. 4.5(b) RC network.

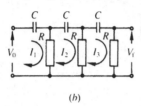

(b)

phase shift to the input signal applied to it and the phase shift network provides another 180° phase shift at the frequency of oscillation thus fulfilling the phase shift condition for oscillation.

By making all the resistors equal to R and the capacitors to C in the RC network, which has been shown separately in Fig. 4.5(b) for convenience, we may write,

$$\left.\begin{array}{c} \dfrac{1}{j\omega C}I_1 + R(I_1 - I_2) = V_o \\[2mm] R(I_2 - I_1) + \dfrac{1}{j\omega C}I_2 + R(I_2 - I_3) = 0 \\[2mm] R(I_3 - I_2) + \dfrac{1}{j\omega C}I_3 + RI_3 = 0 \\[2mm] -RI_3 = V_f \end{array}\right\} \tag{4.9}$$

Say $\alpha = 1/\omega CR$, then the expressions (4.9) can be rewritten as

$$\left.\begin{array}{c} -j\alpha I_1 + I_1 - I_2 = V_o/R \\ I_2 - I_1 - j\alpha I_2 + I_2 - I_3 = 0 \\ I_3 - I_2 - j\alpha I_3 + I_3 = 0 \\ -I_3 = V_f/R \end{array}\right\} \tag{4.10}$$

From the expressions (4.10) it can be shown that

$$\frac{V_f}{V_o} = \frac{1}{-1 + 5\alpha^2 + j(6\alpha - \alpha^3)} \tag{4.11}$$

The ratio is the feedback fraction β of the RC network. Since the phase shift due to this network should be 180°, β must be real. In other words the imaginary part of the denominator must be equal to zero. So we may say,

$$6\alpha - \alpha^3 = 0$$

or

$$\alpha = \sqrt{6} \tag{4.12}$$

or

$$\omega CR = 1/\sqrt{6}$$

$$\therefore \qquad f = \frac{1}{2\pi\sqrt{6}RC} \qquad\qquad (4.13)$$

Now substituting the value of α from (4.12) into (4.11),

$$\beta = V_f/V_o = 1/(-1 + 5 \times 6) = 1/29 \qquad\qquad (4.14)$$

In order to satisfy the magnitude condition for oscillation $|A\beta| = 1$, it is therefore required that $|A|$ should be at least 29.

Worked example 4.4

Design an *RC* phase-shift oscillator to operate at a frequency of 250 Hz.

Solution
From (4.13)

$$f = 1/2\pi\sqrt{6}RC$$

Let C be 1 μF, then $R = 1/2\pi f \times 10^{-6} \times \sqrt{6} = 259\,\Omega$.

*Now from (4.14) $\beta = 1/29$, so the gain of the forward amplifier has to be at least 29. Let us use an operational amplifier as the active device instead of the transistor as shown in Fig. ex.4.4. The gain is given by,

$$R_2/R_1 = 29$$

If we choose R_1 to be 10 kΩ to prevent R_1 from loading the R of the *RC* network then $R_2 = 29 \times 10 \times 10^3 = 290$ kΩ.

Fig. ex.4.4

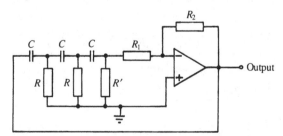

*   Since 259 Ω would cause high currents from the op.amp. let us use 100 nF and 2.59 kΩ for C and R respectively.

The value of the last resistor of the RC network has to be recalculated so that this in parallel with R_1 gives the required value of $2.59\,\text{k}\Omega$ for exact oscillation frequency. Say it is R' then

$$\frac{R' \times 10 \times 10^3}{R' + 10 \times 10^3} = 2.59 \times 10^3$$

therefore

$$R' = 3.5\,\text{k}\Omega$$

4.4   Wien-bridge oscillator

The Wien-bridge oscillator is another type of RC feedback oscillator. It is very popular and most often uses an operational amplifier for the forward amplification as shown in Fig. 4.6. Resistors R_1 and R_2 connected to the inverting input terminal of the operational amplifier determine the gain of the amplifier and are selected so that the magnitude of the loop gain is unity. The parallel combination of C_A and R_A and the series combination of C_B and R_B form the feedback network which is connected to the noninverting terminal of the operational amplifier. This positive feedback gives rise to the oscillation. In order to determine the frequency of oscillation let us first find the feedback fraction β. From Fig. 4.6 we may write

$$\beta = Z_A/(Z_A + Z_B) \tag{4.15}$$

$$= \frac{R_A \times \dfrac{1}{j\omega C_A}}{R_A + \dfrac{1}{j\omega C_A}} \Bigg/ \frac{R_A \times \dfrac{1}{j\omega C_A}}{R_A + \dfrac{1}{j\omega C_A}} + R_B + \frac{1}{j\omega C_B}$$

$$= \frac{R_A}{1 + j\omega C_A R_A} \Bigg/ \left(\frac{R_A}{1 + j\omega C_A R_A} + R_B + \frac{1}{j\omega C_B} \right)$$

$$= \frac{j\omega R_A C_B}{j\omega C_B R_A + (1 + j\omega C_A R_A) j\omega C_B R_B + (1 + j\omega C_A R_A)}$$

Fig. 4.6. The Wien-bridge oscillator.

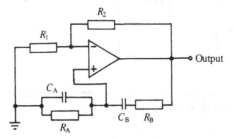

$$= \frac{\omega R_A C_B}{\omega C_B R_A + \omega C_B R_B + \omega C_A R_A + j(\omega^2 C_A R_A C_B R_B - 1)} \quad (4.16)$$

In order to meet the phase shift condition for oscillation β should be a real quantity. Therefore the imaginary part of the denominator of (4.16) must be zero. So we have

$$\omega^2 C_A C_B R_A R_B - 1 = 0$$

or

$$\omega^2 = \frac{1}{C_A C_B R_A R_B}$$

In practice we make $R_A = R_B = R$ and $C_A = C_B = C$.

Thus the frequency of oscillation becomes

$$f = 1/2\pi RC \quad (4.17)$$

Substituting the value of f from (4.17) into (4.16) we get

$$\beta = 1/3$$

Another condition to allow oscillation to occur is to make the magnitude of the loop gain unity, i.e. $|A\beta| = 1$. This can be achieved by making

$$A = 1 + R_2/R_1 = 3$$

or

$$R_2 = 2R_1 \quad (4.18)$$

Continuous variation of frequency can be achieved by varying either one of the capacitors only or both simultaneously. The frequency range is normally changed by switching in different values for the two identical resistors R.

Worked example 4.5

Design a Wien bridge-oscillator that oscillates at 25 kHz.

Solution

Let $C_A = C_B = 1$ nF, and $R_A = R_B = R$. Then,

$$f = 25 \times 10^3 = \frac{1}{2\pi R \times 1 \times 10^{-9}}$$

$$\therefore \quad R = \frac{1}{2\pi \times 1 \times 10^{-9} \times 25 \times 10^3} = 6.36 \, k\Omega$$

Let $R_1 = 10 \, k\Omega$. Then from (4.18),

$$R_2 = 2R_1 = 2 \times 10 \times 10^3 = 20 \, k\Omega$$

Note: Because of the component tolerances R_A is not exactly equal to R_B and similarly C_A is not exactly equal to C_B. Therefore R_f should be made adjustable so that the loop gain can be set as necessary to sustain oscillation.

Worked example 4.6

Design a Wien-bridge oscillator which would be able to operate over a frequency range from 500 Hz to 50 kHz.

Solution
From (4.17),

$$f = \frac{1}{2\pi RC}$$

or

$$R = \frac{1}{2\pi f C}$$

Using the upper value of a single turn trimmer capacitor of swing 6.5 pF to 65 pF, for the lowest frequency, i.e., 500 Hz,

$$R = \frac{1}{2\pi \times 500 \times 65 \times 10^{-12}} = 4.9 \, \text{M}\Omega$$

When the capacitor is set at 6.5 pF the oscillator frequency will be 5 kHz. Now for the frequency range of 5 kHz to 50 kHz,

$$R = \frac{1}{2\pi \times 5000 \times 65 \times 10^{-12}} = 490 \, \text{k}\Omega$$

At this value of R, when the capacitor is set at 6.5 pF, the oscillator frequency will be 50 kHz.

Fig. ex.4.6

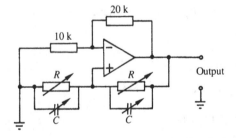

Now for the gain assume

$$R_1 = 10\,\text{k}\Omega$$

Then from (4.18),

$$R_2 = 2 \times 10 \times 10^3 = 20\,\text{k}\Omega$$

For a closed loop gain of 3, at 50 kHz, an operational amplifier of the 741 type may be used.

Rs and Cs are ganged resistors and capacitors. R is switched from 4.9 MΩ to 490 kΩ to change from one frequency range to another and C is varied continuously from 6.5 pF to 65 pF to cover all the frequencies in these ranges.

Amplitude stabilization of Wien-bridge oscillators

A modified Wien-bridge oscillator is shown in Fig. 4.7. This provides a stable sinusoidal output, independent of any change in supply voltage. The resistor R_1 of the circuit of Fig. 4.6 is replaced by the FET Q_1. The output is fed back to the gate of the FET after rectification and smoothing by the diode and the combination of R and C respectively. The dynamic resistance of the FET varies linearly with V_{GS} for small values of drain-source voltage. Therefore any variation in supply voltage, which apparently changes the output voltage, will change the dynamic resistance of the FET thus keeping the output constant. The frequency of oscillation of the circuit shown in Fig. 4.7 is

$$f = 1/2\pi \times 4.7 \times 10^3 \times 6800 \times 10^{-12} = 5\,\text{kHz}$$

Fig. 4.7. An amplitude stabilized Wien-bridge oscillator.

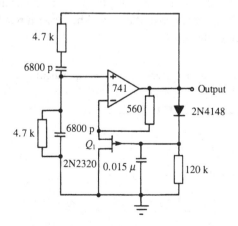

Variation in frequency of the circuit shown in Fig. 4.7 has been found to be only 0.02% with a 30% change in supply voltage.

4.5   Crystal oscillator

A crystal oscillator provides a very stable frequency output. It uses a piece of quartz which looks like frosted glass. It is cut and polished to vibrate at a certain frequency. Quartz has a piezoelectric effect, i.e. a strain generates a low voltage output at the surface of the crystal. Conversely if a voltage is applied to the quartz, it will be distorted physically. Because of this when a quartz crystal is properly mounted deformations take place within the crystal and an electromechanical system is formed. If a voltage is now applied to the electrodes plated on to its opposite surfaces then the system will vibrate at a specific resonant frequency. The resonant frequency and the Q ($f_o/\Delta f_{3\text{-dB}}$, quality factor) depend on (a) the crystal dimension, (b) the orientation of the crystal surfaces with respect to its axes and (c) the method of mounting the crystal. A typical frequency range is 10 kHz to 30 MHz. Oscillators with very high Qs in the range 1000–100 000 can be obtained using quartz crystals.

Typical crystal length or width is 15–30 mm. A typical value for the thickness is 4 mm. The frequency of oscillation of a crystal depends on its thickness. The thinner it is the higher the frequency of oscillation.

A crystal can be represented as shown in Fig. 4.8. L, C and R are the analogue equivalents of the mass, the compliance and the viscous damping factor of the mechanical system, and C' represents the electrostatic capacitance between the two electrodes with the crystal as a dielectric. Thus the equivalent circuit consists of two capacitors, giving a pair of closely spaced series and parallel resonant frequencies. R which is of the order of a few kΩ is very small compared with the reactance of the capacitor C and inductor L whose typical values can be 0.05 pF and 3 H respectively. Therefore if we ignore R the impedance of the crystal becomes a reactance and from Fig. 4.8 we may write,

Fig. 4.8. The symbol of a quartz crystal and its equivalent circuit.

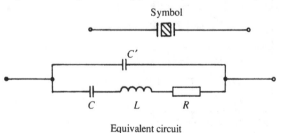

Symbol

Equivalent circuit

$$jX = \frac{\left(j\omega L + \dfrac{1}{j\omega C}\right) \times \dfrac{1}{j\omega C'}}{j\omega L + \dfrac{1}{j\omega C} + \dfrac{1}{j\omega C'}} = -\frac{j}{\omega C'} \frac{\omega^2 - \dfrac{1}{LC}}{\omega^2 - \dfrac{1}{L}\left(\dfrac{1}{C} + \dfrac{1}{C'}\right)}$$

$$= \frac{-j}{2\pi f C'} \cdot \frac{f^2 - \dfrac{1}{4\pi^2 LC}}{f^2 - \dfrac{1}{4\pi^2 L}\left(\dfrac{1}{C} + \dfrac{1}{C'}\right)} = -\frac{j}{2\pi f C'} \cdot \frac{f^2 - f_s^2}{f^2 - f_p^2}$$

The above expression shows that the crystal can have both series and parallel resonance. The series resonant frequency, $f_s = 1/2\pi\sqrt{(LC)}$ and the parallel resonant frequency, $f_p = 1/2\pi\sqrt{\left[\dfrac{1}{L}\left(\dfrac{1}{C} + \dfrac{1}{C'}\right)\right]}$.

Figure 4.9 shows that the reactance changes rapidly with frequency. Since the value of C' is very much greater than C, f_p is approximately equal to f_s. A typical difference between the values of f_p and f_s is 0.3%. In practice a crystal works as a parallel resonant circuit, therefore operates nearer to f_p than f_s.

By examining Fig. 4.9 we can say that the reactance is inductive within this frequency range. A crystal can therefore be used instead of the inductor in a Colpitts oscillator as shown in Fig. 4.10. This circuit known as the Pierce crystal oscillator is very popular since it provides highly stable frequency and uses only a very few components. The circuit only oscillates when the crystal behaves as an inductor and hence the oscillation frequency has to lie somewhere between f_s and f_p. The crystals available on the market have frequencies typically in the range from 32.768 kHz to 20.0 MHz. Their

Fig. 4.9. Reactance versus frequency characteristic of a quartz crystal.

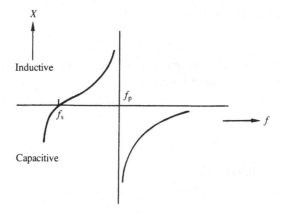

Fig. 4.10. The Pierce crystal oscillator.

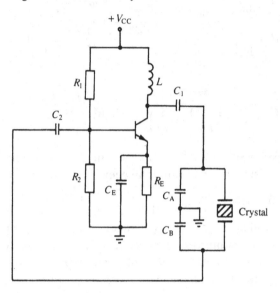

typical tolerance at 25 °C is 20 ppm and temperature stability is 50 ppm within the temperature range of − 10 °C−+ 60 °C. Doublers and triplers are used to double or triple the frequency of a crystal oscillator output. Again digital counters can be used to divide the frequencies in order to obtain lower values of frequencies.

Worked example 4.7
A quartz crystal used in a Colpitts oscillator has the following parameters: $L = 3.3$ H, $C = 0.065$ pF, $R = 10$ kΩ and $C' = 4$ pF. What is the frequency at which the oscillator will operate?

Solution
The resistance R is low compared to the reactance at RF and we may ignore it. The series resonant frequency,

$$f_s = \frac{1}{2\pi\sqrt{(LC)}}$$

$$= \frac{1}{2\pi\sqrt{(3.3 \times 0.065 \times 10^{-12})}}$$

$$= 344 \text{ kHz}$$

The parallel resonant frequency,

$$f_p = \frac{1}{2\pi} \sqrt{\left[\frac{1}{L} \left(\frac{1}{C} + \frac{1}{C'} \right) \right]}$$

$$= \frac{1}{2\pi} \sqrt{\left[\frac{1}{3.3} \left(\frac{1}{0.065 \times 10^{-12}} + \frac{1}{4 \times 10^{-12}} \right) \right]}$$

$$= 347\,\text{kHz}.$$

The oscillator will operate at a frequency within the range 344 to 347 kHz, the percentage difference being $\frac{3}{344} \times 100 = 0.8\%$.

Summary

1. A feedback circuit produces oscillations provided the magnitude of the loop gain is at least unity and the phase shift of the loop gain is of a multiple of 360° at the frequency of oscillation. The conditions are known as the Barkhausen criterion.

2. LC oscillators provide radio frequencies needed for radio receivers and transmitters. They oscillate at the resonant frequency of the LC tuned circuit. Hartley and Colpitts oscillators are popular LC oscillators. In the former one a tapped coil is used in the tuning circuit to provide the feedback voltage from the output phase-shifted by 180°. This phase shift and the phase shift due to the forward amplifier produces a total phase shift of 360° or 0. The frequency range for this type of oscillator is from 100 kHz to a few MHz. In Colpitts oscillators capacitive feedback is used. It operates at low radio frequencies and also in the VHF band up to 300 MHz. The frequency of oscillation is given by $1/[2\pi\sqrt{(LC)}]$.

3. RC phase shift oscillators operate at audio frequencies. An RC ladder network is used in order to obtain 180° phase shift. The frequency of oscillation is $1/[2\pi\sqrt{6}RC]$. The feedback fraction for this circuit is 1/29, therefore, in order to satisfy the magnitude condition of oscillations the gain of the forward amplifier has to be at least 29.

4. A Wien-bridge oscillator can provide a wide range of low frequencies using variable resistors and capacitors in the feedback path. Most often an operational amplifier is used for the forward amplification. The amplitude stabilization of Wien-bridge oscillator outputs has been discussed in this chapter.

5. Very stable frequency is obtained from a crystal oscillator. It uses a quartz crystal which has piezoelectric properties, and oscillates at the same frequency as the frequency of vibration of the quartz crystal. If necessary, frequencies can be increased by using doublers and triplers, or reduced with the help of divider circuits.

Problems

4.1. The gain of the forward amplifier, A in the circuit of Fig. 4.1 is $-10^7/j\omega$. If the circuit oscillates at 1 kHz find the feedback fraction β.

4.2. In the circuit of Fig. 4.1, the gain of the forward amplifier is given by $A = -4 \times 10^7/j\omega$. If the feedback fraction β is $3.2 \times 10^3/(4 \times 10^3 + j\omega)^2$ will the circuit oscillate? If so, what is the frequency of oscillation?

4.3. The Hartley oscillator shown in Fig. 4.3 has $L_M = 3\,\text{mH}$ and $L_N = 35\,\mu\text{H}$. Determine the range of capacitance values for the variable capacitor C for the case where the frequency of oscillation is varied between 47 kHz and 2.5 MHz.

4.4. The Hartley oscillator shown in Fig. 4.3 has a fixed capacitance of 2 pF. What will be the value of centre-tapped inductor to make the circuit oscillate at 3.7 kHz?

4.5. A Colpitts oscillator is designed with $C_M = 4.7\,\text{nF}$ and $C_N = 100\,\text{pF}$ and a variable inductor. Determine the range of inductance values if the frequency of oscillation is to vary between 47 kHz and 2.5 MHz.

4.6. The transistor in the Colpitts oscillator of Fig. 4.4 has a mutual conductance of 3×10^{-3} siemens, a base-emitter resistance of 1 kΩ, a base spreading resistance of 75 Ω and an output impedance equal to 12 kΩ. Given: $R_1//R_2 = 1.4\,\text{k}\Omega$, $L = 2\,\text{mH}$, $C_M = 75\,\text{pF}$ and $C_N = 22\,\text{pF}$, find the frequency at which the circuit oscillates.

4.7. Design an RC phase-shift oscillator incorporating an operational amplifier to operate at a frequency of 1.5 kHz.

4.8. Design a Wien-bridge oscillator incorporating an operational amplifier to operate at 20.5 kHz.

4.9. Design a Wien-bridge oscillator which would be able to oscillate over a range of frequencies from 950 Hz to 20.5 kHz.

4.10. A quartz crystal used in a Colpitts oscillator has the following parameters: $L = 2.5\,\text{H}$, $C = 0.047\,\text{pF}$, $R = 15\,\text{k}\Omega$ and $C' = 6.8\,\text{pF}$. What is the range of frequencies over which the oscillator operates?

5

Phase-locked loops

Objectives
At the end of the study of this chapter a student should be:

1. familiar with three basic functional blocks of a phase-locked loop
2. familiar with the principle of operation of a phase-locked loop as a whole
3. able to determine important PLL parameters such as lock range and acquisition range
4. familiar with typical applications, both in electro-mechanical and telecommunication systems.

The phase-locked loop (PLL) is a very useful and versatile building block in the frequency domain. It is available from manufacturers as a single integrated circuit. It helps to synchronise the output signal of an oscillator with a reference signal in both frequency and phase. While synchronised in frequency, the phase difference between the output signal and the input signal is zero, or very small. It works in much the same way as a general feedback loop which acts in most control systems, e.g., electronic, mechanical, as shown in Fig. 5.1. Here the input is a function of the desired output. If the output is different from the desired value, the mixer produces an error signal which is then amplified and corrects the output. Although its concept has been known since 1932, its application was restricted until the 1960s, when it first became available in an integrated circuit form.

Fig. 5.1. A general feedback loop.

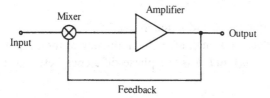

Feedback

Fig. 5.2. Components of phase-locked loops.

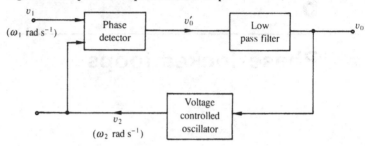

5.1   Components of phase-locked loops

The PLL contains a phase detector, a low-pass filter and a voltage-controlled oscillator in an arrangement as shown in Fig. 5.2. It is similar to the general feedback loop shown in Fig. 5.1. The PLL makes ω_2, the frequency of the output signal v_2 of the voltage-controlled oscillator equal to ω_1, the frequency of the reference signal v_1, by negative feedback under steady state conditions. Note: frequencies will be specified in rad s^{-1} for convenience in this chapter. If the frequencies are not the same then the voltage output of the low-pass filter acts on the voltage-controlled oscillator so that the frequencies become identical. Once the two frequencies are made equal, the error signal from the phase detector becomes a function of the phase difference between the two signals, and the phase difference is then controlled.

5.1.1   Phase detector

Various circuits can be used to build a phase detector. The one we will discuss here is an analogue multiplier that forms the product of two input signals. One of these signals is the reference signal and the other is the output of a limiter circuit into which the output of the voltage-controlled oscillator is fed. Figure 5.3(a) shows the block diagram of the phase detector. Let us consider that the reference signal is given by

$$v_1 = V_1 \sin(\omega_s t + \phi) \tag{5.1}$$

and the output of the voltage-controlled oscillator is given by

$$v_2 = V_2 \sin \omega_s t \tag{5.2}$$

where ω_s is the frequency in rad/s under steady state condition (i.e. the frequency of the reference signal, $\omega_1 =$ the frequency of the output signal from the VCO, $\omega_2 = \omega_s$) and ϕ is the phase difference between the two signals.

Fig. 5.3. (a) Block diagram of a phase detector and (b) the input and output voltage waveforms of a limiter circuit.

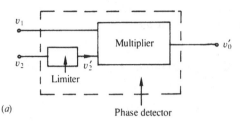

(a)

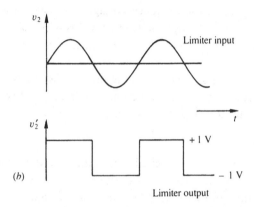

(b)

A comparator with clamping diodes at the output and the reference terminal connected to zero may be used as the limiter circuit yielding an output of $+1$ V when $v_2 > 0$, and -1 V when $v_2 < 0$. Figure 5.3(b) shows the input and output voltages v_2 and v_2' respectively of such a circuit. Thus v_2' is a square wave with an amplitude of 2 V and can be expressed in a Fourier series as follows,

$$v_2' = 2 \times 2/\pi \cdot \sin \omega_s t + 2 \times 2/3\pi \cdot \sin 3\omega_s t + \dots \tag{5.3}$$

Since the phase detector is followed by a low-pass filter which will attenuate all the terms with higher harmonics than the fundamental, for simplicity we may write (5.3) as,

$$v_2' = 4/\pi \cdot \sin \omega_s t \tag{5.4}$$

The reference signal v_1 and the limiter output v_2' are put into the multiplier and thus the output of the phase detector is given by

$$v_0' = K_m v_1 v_2' \tag{5.5}$$

where K_m is the multiplier constant.

Substitution of the values of v_1 and v_2' from (5.1) and (5.4) into (5.5) yields

$$v_0' = K_m V_1 \sin(\omega_s t + \phi) \cdot \frac{4}{\pi} \sin \omega_s t$$

$$= K_m V_1 \frac{2}{\pi} [2 \sin(\omega_s t + \phi) \sin \omega_s t]$$

$$= K_m V_1 \frac{2}{\pi} [\cos \phi - \cos(2\omega_s t + \phi)] \tag{5.6}$$

Therefore if the peak value of the reference voltage is constant then the multiplier output can be made proportional to the cosine function of ϕ provided the higher frequency term $\cos(2\omega_s t + \phi)$ is filtered out with the aid of a low-pass filter.

The conversion gain of a phase detector is the ratio of the change in the magnitude of its output to the change in the phase difference between the two inputs, and therefore, it is expressed in volts rad^{-1}.

5.1.2   Low-pass filter

The performance of a phase-locked loop is influenced not only by the type of phase detector used, but also – to a certain extent – by the type of the low-pass filter. Although there are different kinds of analogue filters available, both active and passive, a simple RC first-order low-pass passive filter as shown in Fig. 5.4 will be considered here because this type is quite suitable for most applications.

The output of the phase detector v_0' is fed to the filter whose output v_0 can now be expressed as

$$v_0 = \frac{\dfrac{1}{j\omega C}}{R + \dfrac{1}{j\omega C}} \cdot v_0' = \frac{1}{1 + j\omega CR} v_0'$$

$$= \frac{v_0'}{\sqrt{\left[1 + \left(\dfrac{\omega}{1/CR} \right)^2 \right]}} = \frac{v_0'}{\sqrt{\left[1 + \left(\dfrac{\omega}{\omega_{LPF}} \right)^2 \right]}} \tag{5.7}$$

Fig. 5.4. A first-order low-pass passive filter.

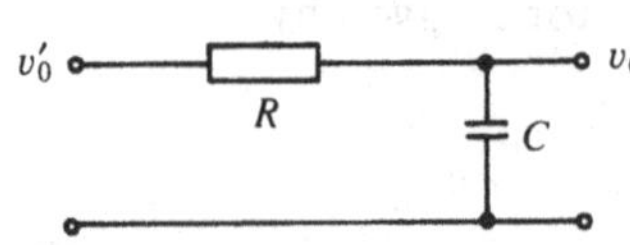

From (5.7) we may observe that the cutoff frequency ω_{LPF} is equal to $1/CR$. The values of C and R should be chosen so that ω_{LPF} is considerably less than $2\omega_s$, thus enabling the filter to attenuate the higher frequency component, $\cos(2\omega_s t + \phi)$ of v_o' in accordance with (5.7). For the low frequency component of the voltage v_o, $\omega = \phi$, and since $\phi \ll \omega_s$ (5.7) yields,

$$v_o = v_o' \tag{5.8}$$

Substituting for v_o' from (5.6) into (5.8) we obtain

$$v_o = K_m V_1 2/\pi \cdot \cos \phi = V_o \cos \phi \tag{5.9}$$

where $V_o = k_m V_1 2/\pi$.

The output of the low-pass filter is therefore a cosine function with a peak value of V_0 which is, in turn, proportional to the input signal amplitude V_1.

5.1.3 Voltage controlled oscillator

The voltage controlled oscillator is an essential part of a phase-locked loop. It may generate either sinusoidal or square waves. Earlier we assumed a sinusoidal function for v_2, the output of the VCO. Here we will consider an astable multivibrator as shown in Fig. 5.5(a) whose frequency of oscillations can be controlled by a d.c. input, in this case, the output of the low-pass filter. In the absence of any input, the oscillator runs at a free-running frequency ω_0 as shown in the dynamic curve of the VCO in Fig. 5.5(b). As the output voltage of the low-pass filter v_o increases, the coupling capacitors charge up more rapidly and the frequency of oscillations ω_2 increases; again with the reduction in the output voltage of the filter ω_2 drops. The VCO works over a limited range and its transfer curve is

Fig. 5.5. (a) A voltage controlled oscillator circuit and (b) its dynamic curve.

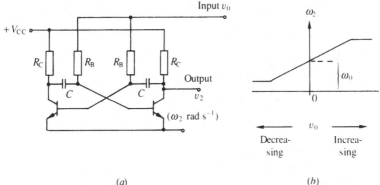

(a)   (b)

approximately linear. We may express the relationship between the oscillatory frequency and the d.c. input to the voltage controlled oscillator as follows:

$$\omega_2 = \omega_0 + K_{vco} v_o \qquad (5.10)$$

where K_{vco} is the gain of conversion of the VCO and its unit is rad V^{-1}. The output of the voltage controlled oscillator, v_2 is fed to the phase detector to complete the loop.

This particular circuit is linear over a small range and therefore, if necessary, a more sophisticated circuit may be used. In integrated circuit form some PLLs include a VCO such as type NE565, but it is common practice to use a separate VCO integrated circuit chip such as type 4024 with a phase detector type 4044 to form a phase-locked loop.

A current controlled oscillator is used in some PLLs instead of the VCO. In these cases the input signal is obtained from a current controlled source rather than a voltage controlled source.

The conversion gain of a VCO is the ratio of the output frequency change to the input magnitude change and is expressed in rad V^{-1}. In the case of current-controlled oscillators the unit is rad amp^{-1}.

5.2   Principles of operation

The plot of the output of a low-pass filter versus the phase difference between the input signal frequency and the oscillator frequency is shown in Fig. 5.6. It can be noted that the low-pass filter output is zero when the phase difference is $+90°$. It can also be observed that the former has its maximum value when there is no phase difference between the signals. Let us consider that the reference signal frequency and the

Fig. 5.6. Filter output vs. phase difference between the input signal frequency and the oscillatory frequency.

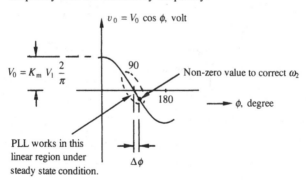

oscillator frequency are perfectly synchronised with the free-running frequency of the VCO. This will make the output of the low-pass filter zero. So we may write

$$v_0 = V_o \cos \phi = 0$$

or,

$$\phi = 90°$$

Hence a PLL normally operates in the region where the phase difference is 90°. This region may be considered to be linear. If the frequency of the reference signal ω_1 now changes from ω_2 then, v_o will adjust and settle to a non-zero value thus making ω_2 equal to the new value of ω_1 and maintaining the frequency lock. It can be observed from the figure that ϕ will be shifted by an amount, $\Delta\phi$, from the reference phase angle of 90°. We may redefine ϕ as $\phi = 90° + \Delta\phi$ for simplicity and the output voltage then becomes

$$v_0 = K_m V_1 \frac{2}{\pi} \cos(90° + \Delta\phi) = -K_m V_1 \frac{2}{\pi}(\sin \Delta\phi) \tag{5.11}$$

Thus v_0 becomes a negative sine function. Figure 5.7 shows the time waveforms of v_0 for small and large input signals. Let us first consider the case when the input signal frequency ω_1 is equal to the VCO output signal frequency ω_2. v_0 is negligible for both small and large input signals when

Fig. 5.7. Time waveforms of filter output for small and large input signals and also small and large phase differences.

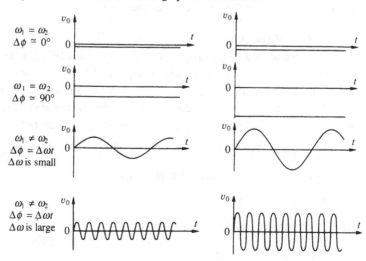

Fig. 5.8. A plot of filter output as a function of phase difference.

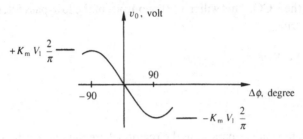

$\Delta\phi \simeq 0$, but, obviously, has a finite d.c. value when $\Delta\phi \neq 0$. The magnitude of the d.c. voltage depends on the input signal magnitude V_1 and $\Delta\phi$.

Now suppose ω_1 is not equal to ω_2, then $\Delta\phi$ is proportional to the difference between the input signal frequency and the oscillator frequency $(\Delta\phi = (\omega_2 - \omega_1)t)$ and v_0 becomes a sinusoidal quantity whose magnitude and frequency depend on V_1 and $\Delta\phi$ respectively.

A plot of v_0 as a function of $\Delta\phi$ is shown in Fig. 5.8. Since a PLL normally operates in the linear region (i.e., around $\Delta\phi = 0°$) we may write,

$$v_0 = \text{slope at } 0° \times \Delta\phi$$

$$= \frac{dv_0}{d\Delta\phi}\Bigg|_{\Delta\phi=0} \times \Delta\phi = \frac{d}{d\Delta\phi}\left(-K_m V_1 \frac{2}{\pi} \sin\Delta\phi\right)\Bigg|_{\Delta\phi=0} \times \Delta\phi$$

$$= -K_m V_1 \frac{2}{\pi}\cos\Delta\phi\Big|_{\Delta\phi=0} \times \Delta\phi = -K_m V_1 \frac{2}{\pi}\Delta\phi \qquad (5.12)$$

Therefore when ω_1 and ω_2 are synchronised, the output of the VCO is proportional to any shift in phase difference.

5.2.1   Lock range (ω_L)

The useful frequency range over which a PLL can track an input signal is called the lock range or tracking range. A PLL can only stay locked onto an incoming signal of frequency ω_1 over a finite range, as shown in Fig. 5.8, i.e., from $\Delta\phi = -90°$ to $\Delta\phi = +90°$. It can be observed from the illustration that at $\Delta\phi = -90°$, v_0 is maximum and is given by

$$v_{0(\text{max})} = K_m 2V_1/\pi \qquad (5.13)$$

and at $\Delta\phi = 90°$, v_0 is minimum and given by

$$v_{0(\text{min})} = -K_m 2V_1/\pi. \qquad (5.14)$$

Under steady-state or locked condition

$$\omega_1 = \omega_2 = \omega_0 + K_{\text{VCO}}v_0 \qquad (5.15)$$

By observing (5.13) and (5.15) we may say that the maximum value of the frequency an input signal may have in order to stay in the locked condition, is given by

$$\omega_{1(\text{max})} = \omega_0 + K_{\text{VCO}}v_{0(\text{max})} = \omega_0 + K_{\text{VCO}}K_m 2V_1/\pi \qquad (5.16)$$

Similarly, the minimum frequency an incoming signal may have in order to stay in the locked condition, is given by

$$\omega_{1(\text{min})} = \omega_0 + K_{\text{VCO}}v_{0(\text{min})} = \omega_0 - K_{\text{VCO}}K_m 2V_1/\pi \qquad (5.17)$$

Therefore, according to the definition, the lock range of the PLL is,

$$\omega_L = \omega_{1(\text{max})} - \omega_{1(\text{min})} = 2K_{\text{VCO}}K_m 2V_1/\pi = \frac{4}{\pi} \cdot K_{\text{VCO}}K_m V_1 \qquad (5.18)$$

By observing (5.18) we can say that the lock range is determined primarily by the maximum frequency swing possible in the voltage-controlled oscillator.

5.2.2 Acquisition range

The frequency range over which a PLL can acquire an input signal is denoted as the acquisition range or capture range. It is limited to a value less than the lock range. Let us consider an input signal whose frequency ω_1 varies with time as shown in Fig. 5.9. The oscillator frequency tracks the incoming signal frequency as long as the latter is within the lock range ω_L, however, as soon as the incoming signal frequency drifts outside the lock range the oscillator frequency ω_2 fails to track it further and settles down to ω_0, the free-running frequency of the oscillator. Thus the PLL becomes unlocked and we may consider that the link between the VCO and the LPF

Fig. 5.9. Operation of a phase-locked loop over a period of time during which the input signal frequency varies significantly.

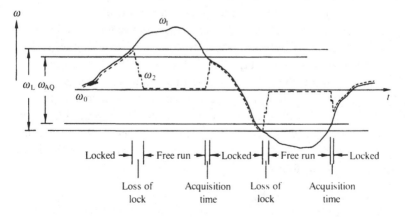

Fig. 5.10. The phase-locked loop may be assumed to be broken when the input signal frequency is outside the lock range.

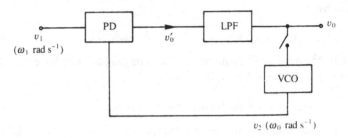

is broken as shown in Fig. 5.10. The output of the phase detector due to the input signal, $V_1 \sin \omega_1 t$, and the oscillator signal, $4/\pi \sin \omega_0 t$, is then given by

$$v_0' = K_m V_1 \sin \omega_1 t \cdot \frac{4}{\pi} \sin \omega_o t$$

$$= K_m V_1 (2/\pi) \{ \cos(\omega_1 - \omega_0)t - \cos(\omega_1 + \omega_0)t \} \qquad (5.19)$$

When the signal v_0' passes through the LPF the higher frequency term is attenuated yielding,

$$v_0 = K_m V_1 (2/\pi) \cos (\omega_1 - \omega_0)t = K_m V_1 (2/\pi) \cos \Delta\omega t \qquad (5.20)$$

where $\omega_1 - \omega_0 = \Delta\omega$.

Since $\omega_1 \neq \omega_2$, ϕ does not exist.

Assuming that the LPF has a cutoff frequency ω_{LPF} we may write the peak value of v_0 as,

$$V_{op} = K_m V_1 2/\pi \frac{1}{\sqrt{[1 + (\Delta\omega/\omega_{LPF})^2]}} \qquad (5.21)$$

Now let us consider that the input signal frequency drifts back towards the lock range. As it approaches the lock range $\Delta\omega$ becomes smaller thus making V_{op} larger. The output of the LPF sinusoidally modulate (frequency modulation) the VCO output according to (5.10). The acquisition begins when the input signal frequency becomes equal to either $\omega_{2(max)}$, in the case of drifting back from outside the upper limit of the lock range, or $\omega_{2(min)}$, in the case of drifting back from outside the lower limit of the lock range. We may now rightly assume that the link between the VCO and the LPF has been reestablished. By using 5.10 we may find the expressions for $\omega_{2(max)}$ and $\omega_{2(min)}$ as follows,

$$\omega_{2(max)} = \omega_0 + K_{VCO} V_{op}$$

and

$$\omega_{2(min)} = \omega_0 - K_{VCO}V_{op} \tag{5.22}$$

The difference between these two limits of the oscillator frequency must give the acquisition range which can be expressed as,

$$\omega_{AQ} = \omega_{2(max)} - \omega_{2(min)} = 2K_{VCO}V_{op}$$

$$= \frac{4}{\pi} K_{VCO}K_m V_1 \frac{1}{\sqrt{\left[1 + \left(\dfrac{\Delta\omega}{\omega_{LPF}}\right)^2\right]}} \tag{5.23}$$

Since $\dfrac{4}{\pi} K_{VCO}K_m V_1 = \omega_L$ and $\Delta\omega = \omega_{AQ}/2$ when the acquisition begins, (5.22) can be rewritten as

$$\omega_{AQ} = \frac{\omega_L}{\sqrt{\left[1 + \left(\dfrac{\omega_{AQ}}{2\omega_{LPF}}\right)^2\right]}} \tag{5.24}$$

Equation (5.24) shows that the acquisition range for a particular lock range is set by the LPF cutoff frequency $\omega_{LPF} = 1/CR$. As the output of the LPF brings the oscillator frequency closer to the input signal frequency, the error signal frequency at the output of the phase detector varies more slowly as shown in Fig. 5.11 and we have asymmetrical error waveforms during acquisition. It produces a non-zero average, i.e. a d.c. component at the LPF output which gradually shifts the oscillator frequency toward the input signal frequency until the PLL is locked. The time required for the PLL to capture the input signal is known as the acquisition or pull-in time as shown in Fig. 5.9.

Fig. 5.11. Phase detector output as the input signal frequency traverses from outside the lock range to inside the acquisition range.

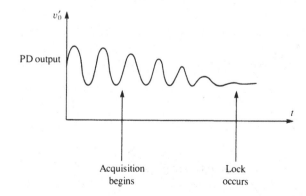

5.2.3 'No-lock' conditions and remedies

A lock cannot be established if the input signal magnitude is too small to drive the VCO in order to produce the necessary deviation in the oscillation frequency, or the input signal frequency is beyond the dynamic range of the VCO. In order to obtain a lock, in the former case, V_1 should be increased internally by employing an additional amplifier in the loop preferably after the LPF or, by adjusting upward the response of the LPF. An amplifier external to the loop may also be used to increase the magnitude of V_1. Next, considering frequency, the free-running frequency of the VCO should be shifted closer to the input signal frequency by adjusting the values of components of the VCO so that the maximum possible frequency swing of the voltage controlled oscillator can embrace the input signal frequency.

Worked example 5.1

A phase-locked loop comprises (a) a phase detector with an output voltage equal to $0.0318[\cos\phi - \cos(2\omega t + \phi)]$ for an input signal having a magnitude of 0.05 V, (b) a simple low-pass filter having a cutoff frequency of 750 Hz and (c) a voltage controlled oscillator with a free-running frequency of 1.5×10^5 Hz and a conversion gain of 1.5×10^4 HzV^{-1}. Calculate (i) the lock range, (ii) the acquisition range and (iii) the steady-state phase error if the input signal has a frequency of 1.5015×10^5 Hz.

Solution

Comparing (5.6) with the given output voltage of the phase detector, we get,

$$K_m V_1 \frac{2}{\pi} = 0.0318$$

or

$$K_m = \frac{0.0318 \times \pi}{2 \times 0.05} = 1$$

From (5.18)

$$\omega_L = \frac{4}{\pi} K_m K_{VCO} V_1$$

$$= \frac{4}{\pi} \times 1 \times 1.5 \times 10^4 \times 2\pi \times 0.05 = 6000 \text{ rad s}^{-1}$$

From (5.24)

$$\omega_{AQ} = \frac{\omega_L}{\sqrt{\left[1 + \left(\dfrac{\omega_{AQ}}{2\omega_{LPF}}\right)^2\right]}}$$

or

$$\omega_{AQ}{}^4 + 4\omega_{LPF}{}^2\omega_{AQ}{}^2 - 4\omega_{LPF}{}^2\omega_L{}^2 = 0$$

The above is a quadratic equation for $\omega_{AQ}{}^2$. So solving for $\omega_{AQ}{}^2$ yields,

$$\omega_{AQ}{}^2 = \frac{-4\omega_{LPF}{}^2 \pm \sqrt{(16\omega_{LPF}{}^4 + 4 \times 4 \times \omega_{LPF}{}^2\omega_L{}^2)}}{2}$$

$$= -4(750 \times 2\pi)^2 \pm \sqrt{16(750 \times 2\pi)^4}$$

$$\frac{+4 \times 4 \times (750 \times 2\pi)^2 \times (6000)^2}{2}$$

$$= +27.8 \times 10^6 \text{ or } -116 \times 10^6$$

taking the positive value for the solution,

$$\omega_{AQ} = 5273 \text{ rad s}^{-1}$$

From (5.11) the steady state error

$$\Delta\phi = \sin^{-1}\left(-\frac{\pi}{2K_m V_1}v_0\right) \tag{a}$$

but

$$v_0 = \frac{\omega_2 - \omega_0}{K_{VCO}}$$

from (5.10). At steady state

$$\omega_1 = \omega_2$$

so

$$v_0 = \frac{\omega_1 - \omega_0}{K_{VCO}} = \frac{2\pi(1.5015 - 1.5) \times 10^5}{2\pi \times 1.5 \times 10^4}$$

$$= 0.01 \text{ volt}$$

Substituting this value into (a) above

$$\Delta\phi = \sin^{-1}\left(-\frac{\pi \times 0.01}{2 \times 1 \times 0.05}\right) = -18°$$

Worked example 5.2

Data for the phase-locked loop in Fig. 5.2 is:

$$K_m = \frac{\pi}{2}, \quad \omega_0 = 10^6 \text{ rad s}^{-1}$$

$$\omega_1 = 1.1 \times 10^6 \text{ rad s}^{-1}, \quad \omega_{LPF} = 10^5 \text{ rad s}^{-1}$$

$$K_{VCO} = 10^5 \text{ rad s}^{-1} \text{ V}^{-1}$$

(i) The amplitude of the input signal V_1 is relatively strong initially, then it gradually fades to zero, and after a while it returns to full strength again. At what values of V_1 is lock lost and then regained?

(ii) Under steady state, calculate v_0 and the phase error if the input signal has a magnitude of 10 V.

Solution

Figure 5.9 shows that the lock is lost when $\omega_1 - \omega_0$ is about to become larger than $\frac{\omega_L}{2}$, i.e., from the point of view of V_1, when V_1 is small enough to make $\omega_1 - \omega_0 = \frac{\omega_L}{2}$.

From (5.18), $\omega_L = \frac{4}{\pi} K_m K_{VCO} V_1$ or

$$V_1 = \frac{\pi \omega_L}{4 K_m K_{VCO}} = \frac{\pi \times 2(\omega_1 - \omega_0)}{4 K_m K_{VCO}}$$

$$= \frac{\pi \times 2(1.1 - 1) \times 10^6}{4 \times \pi/2 \times 10^5}$$

$$= 1 \text{ volt}$$

Therefore at $V_1 = 1$ volt, the lock is lost.

Again Fig. 5.9 shows that the lock is reacquired when V_1 becomes large enough to make

$$\omega_1 - \omega_0 = \frac{\omega_{AQ}}{2}$$

or

$$\omega_{AQ} = 2(\omega_1 - \omega_0) = 2(1.1 - 1) \times 10^6$$

$$= 2 \times 10^5 \text{ rad s}^{-1}$$

but according to (5.24)

$$\omega_{AQ} = \frac{\omega_L}{\sqrt{1 + \left(\dfrac{\omega_{AQ}}{2\omega_{LPF}}\right)^2}} = \frac{\omega_L}{\sqrt{1 + \left(\dfrac{2 \times 10^5}{2 \times 10^5}\right)^2}} = \frac{\omega_L}{\sqrt{2}}$$

so

$$\omega_L = \sqrt{2}\,\omega_{AQ} = \sqrt{2} \times 2 \times 10^5 = 2.8 \times 10^5 \text{ rad s}^{-1}$$

Now with the help of (5.18) we can find the corresponding input voltage as follows:

$$V_1 = \frac{\pi\omega_L}{4K_m K_{VCO}} = \frac{\pi \times 2.8 \times 10^5}{4 \times \pi/2 \times 10^5}$$

$$= 1.414 \text{ volts}$$

Therefore the lock is regained when V_1 becomes equal to 1.414 volts. Under steady state

$$\omega_1 = \omega_2 = \omega_0 + K_{VCO}v_0$$

or

$$v_0 = \frac{\omega_1 - \omega_0}{K_{VCO}} = \frac{(1.1 - 1) \times 10^6}{10^5} = 1 \text{ volt}$$

From (5.11) the steady state error

$$\Delta\phi = \sin^{-1} - \frac{\pi v_0}{2K_m V_1}$$

Now substituting for v_0 we have

$$\Delta\phi = \sin^{-1} - \frac{\pi \times 1}{2 \times \pi/2 \times 10} = \sin^{-1} - \frac{1}{10} = -5.7°$$

5.3 Applications
DC motor speed control

Many electro-mechanical systems, such as magnetic tape drives, require precise speed control, particularly during start and stop operations. This can be achieved by incorporating the motor within a phase-locked loop as shown in Fig. 5.12. When the motor rotates, the tachogenerator produces an a.c. signal whose frequency, ω_2 is proportional to the speed of the motor. This signal is then fed to the phase detector together with the control signal with a frequency ω_1 where ω_1 is proportional to the desired speed of the motor. The phase-locked loop detects any difference between

Fig. 5.12. Control of motor speed with the help of a phase locked loop.

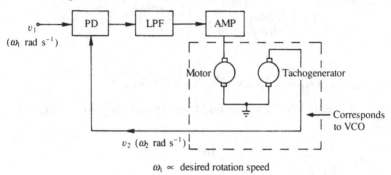

ω_1 ∝ desired rotation speed

ω_1 and ω_2 and drives the motor so that there is a cycle-for-cycle correlation between the control frequency and the speed of the motor.

A slotted opto-switch comprising an infra-red source and sensor may be used in order to obtain an a.c. signal proportional to the speed of the motor. In this case a low-inertia disc with evenly spaced holes along its circumference is mounted on the shaft of the motor as shown in Fig. 5.13. The switch is so placed that the edge of the disc passes through the slots and the holes are aligned with the path of the infra-red beam. As the disc rotates the disc–holes combination alternately breaks and transmits the infra-red beam, and produces a train of pulses whose frequency is directly proportional to the speed of the motor. A Schmitt trigger is commonly used after this stage to obtain a clean square wave, the frequency of which is then compared with the desired frequency of the input signal in the phase detector. Any difference between these frequencies gives an error signal which after filtering yields a d.c. voltage. The latter is amplified to boost the power for the motor so that the speed can be corrected. Eventually the system settles at zero-velocity error.

The accuracy of this system depends mainly on the quality of the device which converts the speed of the motor into an a.c. signal. With the help of a microprocessor and a suitable program a system can be set up so that a

Fig. 5.13. A slotted opto-switch.

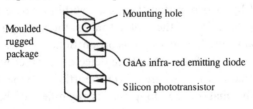

Fig. 5.14. Block diagram of a frequency synthesizer.

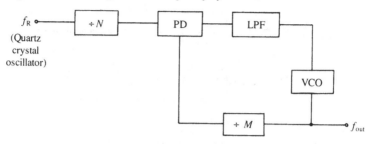

motor can be run at several speeds during different time intervals over a period of time.

Frequency synthesizer

Frequency synthesizer is another important application of the PLL. Discrete frequencies can be generated from a single stable source, these being fractional multiples of the frequency of the source. Figure 5.14 shows a block diagram of a frequency synthesizer. Here a quartz crystal osillator has been used as the reference source. Its frequency is divided by N with the help of a divide-by-counter, before the signal is fed to the phase detector. It is compared with the fractional multiple of the output signal frequency which is obtained by dividing the output frequency by another divide-by-counter. In this case the output of the VCO is the output of the synthesizer. Under steady state conditions the frequencies of the two inputs to the phase detectors must be the same, i.e.

$$\frac{f_R}{N} = \frac{f_{out}}{M}$$

or,

$$f_{out} = \frac{M}{N}f_R$$

we may obtain various frequencies by varying N in powers of 10 and M in steps of 1. In order to obtain a frequency multiplier (integral), we set $N = 1$, in other words, we feed the input straight to the phase detector. With the help of a computer the values of M and N can be chosen and using programmable counters any of a large number of frequencies fractionally related to f_R can be synthesized, each having the stability of the single reference source. Frequency synthesizers are used in FM radios and TV receivers. Crystal oscillators, in the range of 1–10 MHz are commonly used as the reference source.

Fig. 5.15. Block diagram of a frequency modulator.

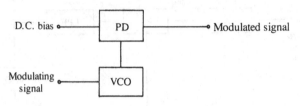

Frequency modulator and demodulator

The PLL is inherently a frequency modulator or demodulator. In order to generate a frequency modulated signal, part of a phase-locked loop may be used. In this case, the free-running frequency of the voltage controlled oscillator is the carrier frequency. The modulating signal is encoded on to the carrier as shown in Fig. 5.15. Here the VCO acts as a frequency modulator and the phase detector as a buffer amplifier. The value of the d.c. bias determines the gain of the amplifier.

One of the methods of recovering the modulating signal from the modulated signal using a phase-locked loop is shown in Fig. 5.16. Here the PLL is brought into lock with the input signal according to the PLL dynamics already discussed. Thus the control voltage of the VCO is the desired demodulated signal. The cutoff frequency of the LPF should be close to the range of the frequency of the demodulated signal.

Fig. 5.16. Block diagram of a frequency demodulator.

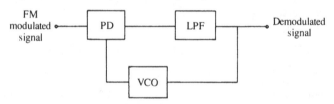

AM detector

A synchronous amplitude-modulator detector can be designed using phase-locked loops as illustrated in Fig. 5.17. The synchronous detection needs a reference signal whose frequency and phase are identical to those of the amplitude modulated signal. The first PLL is used to obtain the required reference signal. It locks onto the carrier frequency of the modulated signal. Its output has an inherent 90° phase-shift with respect to

Fig. 5.17. Block diagram of an amplitude modulation detector circuit.

```
                    ┌──────────────┐
                    │     90°      │
                    │ phase shifter│
                    └──────────────┘
     ┌───────────────────────────────────────────────────┐
     │  ┌─────┐   ┌─────┐   ┌─────┐  ║  ┌─────┐   ┌─────┐  │
─────┤  │ PD  ├───┤ LPF ├───┤ VCO ├──║──┤ PD  ├───┤ LPF ├──┤────
AM   │  └─────┘   └─────┘   └─────┘  ║  └─────┘   └─────┘  │   Demodulated
signal│                              ║     └───────────────┘   signal
     │         └──────────────────────        Part of PLL 2
     └───────────────────────────────┘
                   PLL 1
```

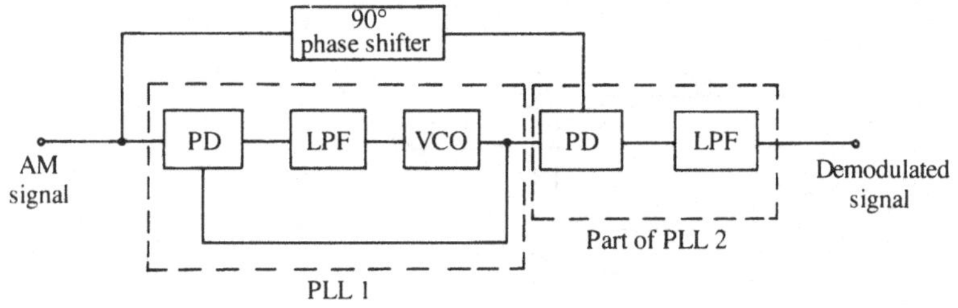

the input as discussed in the theory of PLLs in Section 5.2. In order to ensure that both inputs to the phase detector of the second PLL are in phase, so that no phase errors occur, an intentional 90° phase-shift is provided to the AM input before being inputted to the second PLL. The phase detector of the second PLL multiplies the two inputs. The low-pass filter removes the sum frequency component from the output of the phase detector and the difference frequency component remains as the demodulated AM signal. Thus a signal directly proportional to the amplitude of the input signal is obtained as shown in Fig. 5.18. Note that only the phase detector and the LPF of the second PLL has been used and the loop need not be completed. AM detection with PLLs offer a higher degree of noise immunity than can be obtained by using conventional peak detector type AM detectors.

The RS NE565 Phase-Locked Loop is a self-contained, adaptable filter and demodulator for the frequency range from 0.001 Hz to 500 kHz. The circuit comprises a phase comparator, an amplifier, a low-pass filter and a

Fig. 5.18. Time waveforms at various stages of the amplitude modulation detector circuit.

90° phase-
shifted AM
wave

Output of
PLL 1

Output of
2nd PD

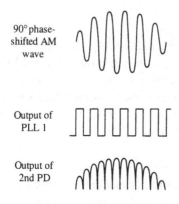

Fig. 5.19. Block diagram of phase locked loop type RS NE565. (RS Components Ltd.)

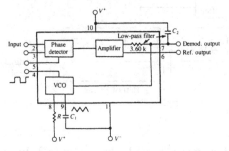

Fig. 5.20. VCO characteristics of the PLL type RS NE565. (RS Components Ltd.)

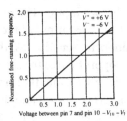

Fig. 5.21. Lock range as a function of input voltage for the PLL type RS NE565. (RS Components Ltd.)

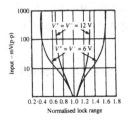

voltage-controlled oscillator of exceptional stability and linearity as shown in schematic form in Fig. 5.19. The free-running frequency of the VCO can be adjusted externally with a resistor or a capacitor. The low-pass filter, which determines the acquisition characteristics of the PLL, is formed by an internal resistor and an external capacitor. Typical performance characteristics are shown in Figs. 5.20 and 5.21.

Summary

1. Phase-locked loops are available in integrated circuit form and are used as versatile building blocks in the frequency-domain.

2. The main elements of a phase-locked loop are the phase detector, the low-pass filter and the voltage controlled oscillator. If necessary an amplifier may be included in the loop.

3. The phase-locked loop forces the frequency of the voltage controlled oscillator to be synchronous with the input signal frequency. Under this (steady state) condition the output of the low-pass filter is proportional to the phase difference between the input signal and the VCO output and thus the phase is controlled.

4. The frequency range over which a PLL can track an input signal is called the lock or tracking range. This range is determined by the maximum frequency swing possible in the voltage controlled oscillator.

5. The range of input signal frequencies over which the PLL can acquire a lock from an unlocked condition is somewhat smaller than the lock range. The time required to acquire a signal is known as pull-in or acquisition time.

6. A lock cannot be established if (a) the input signal magnitude is too small to produce the necessary deviation in the VCO frequency, or (b) the input signal frequency is beyond the dynamic range of the VCO. In order to achieve a lock, an amplification of the input signal is required in the case of (a), and the free-running frequency of the VCO should be shifted closer to the input signal frequency in the case of (b) by adjusting the components of the VCO.

7. Major applications of PLLs are in the fields of motor speed control and telecommunications.

Problems

5.1. A sinusoidal signal having a magnitude of 5 volt peak is fed into a phase-locked loop comprising a multiplier with a constant of one and a voltage controlled oscillator having a conversion gain of 10^4 rad s^{-1} V^{-1}. Find the lock range.

5.2. A phase-locked loop comprises (a) a multiplier with a constant of 1.54, (b) a voltage controlled oscillator which has a frequency–voltage characteristic given by $\omega_2 = \omega_0 + 1.1 \times 10^3 v_0$ rad s^{-1} and (c) a low-pass filter. Find the lock range for the phase-locked loop if the input voltage is 1 volt.

5.3. If the low-pass filter of the phase-locked loop of Problem 5.1 has a characteristic

$$G(j\omega) = \cfrac{1}{1 + \cfrac{j\omega}{1500}}$$

find the acquisition range of the phase-locked loop.

5.4. The multiplier of the phase-locked loop shown in Fig. 5.2 has a constant of 0.95 and the voltage controlled oscillator has a characteristic $\omega_2 = 10^6 + 10^5 v_0$ rad s^{-1}. The phase-locked loop is under steady state condition when a signal with a frequency of 1.1×10^6 rad s^{-1} is fed to it. Suppose the amplitude of the signal gradually fades to zero. Determine the amplitude of the signal at which the lock-in is lost.

5.5. The low-pass filter of the phase-locked loop of Problem 5.1 has a characteristic

$$G(j\omega) = \frac{1}{1 + \dfrac{j\omega}{10^5}}$$

If the magnitude of the input signal is gradually decreased, the lock is lost. Find the magnitude of the signal at which the lock would be regained?

5.6. Calculate the phase-error under steady state condition for the phase locked loop of Problem 5.4, when the input signal is $5 \sin 1.1 \times 10^6 t$.

5.7. A phase-locked loop comprises (a) a phase detector with output voltage equal to $0.0636[\cos \phi - \cos(2\omega t + \phi)]$ for an input signal having a magnitude of 0.01 volt, (b) a low-pass filter with a cutoff frequency of 1 kHz and (c) a voltage controlled oscillator with a free-running frequency of 10^5 Hz and conversion gain of 10^4 Hz V^{-1}. Calculate (i) the lock range, (ii) the acquisition range and (iii) the steady state phase-error if the input signal has a frequency of 1.0015×10^5 Hz.

5.8. Examine whether an input signal with a frequency of 10^5 Hz can be tracked by a phase-locked loop comprising a voltage controlled oscillator which has a frequency–voltage characteristic given by

$$\omega_2 = 10^4 + 10^3 v_0 \text{ rad s}^{-1}$$

The output of the low pass filter in this case is 4.75 volt.

5.9. For the phase-locked loop shown in Fig. P.5.9, the input voltage has a peak value of 0.5 volt and the oscillator has a peak voltage of 1 volt. The

Fig. P.5.9

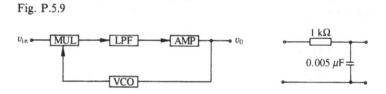

amplifier gain is 1000. The frequency–voltage characteristic of the voltage controlled oscillator is $\omega_2 = \omega_0 + 200v_0$. The low-pass filter is of the form shown in the figure. Determine the lock range and the acquisition range.

5.10. Design a phase-locked loop whose acquisition range is 9 kHz and lock range is 10 kHz.

6

Modulation in communication systems

Objectives

At the end of the study of this chapter students should be:

1. familiar with radio frequency carrier waves and low frequency information signals
2. familiar with the principles of different types of modulation
3. able to choose a suitable method of modulation for transmission of a particular signal.

In a communication system radio frequencies are used to carry information which is most often of low frequencies from one place to another mainly because of the ease with which the high frequency waves can propagate around the world by multiple reflections from the ionosphere and at very high frequencies antennae of modest size can form narrow beams. Ranging from roughly 3 kHz to 300 GHz radio frequencies are widely used in telephone systems, radio and television broadcasting, satellite communications and also in radio detection and ranging. Usually the information is at audio frequency and is transposed onto the radio frequency to be carried from one point to another. The process of transposition is known as modulation. There are several ways to modulate a radio wave. In this chapter we will discuss different types of modulators and also demodulators. The latter are used to recover information from modulated waves at the receiving point.

6.1   Amplitude modulation

In this type of commonly used modulator a sinusoidal carrier wave is made to vary in amplitude in sympathy with the magnitude of the low frequency information signal. The information signal is also called the baseband signal or the modulating signal. Let us consider that a simple sinusoid carrier wave $c(t)$ of a constant amplitude A_c is modulated by a composite baseband signal $m(t)$. The carrier wave and the baseband signal are expressed as follows,

$$c(t) = A_c \cos \omega_c t \qquad (6.1)$$

Fig. 6.1. Amplitude modulation.

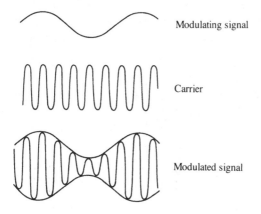

Modulating signal

Carrier

Modulated signal

and

$$m(t) = B_1 \cos \omega_1 t + B_2 \cos \omega_2 t + B_3 \cos \omega_3 t \qquad (6.2)$$

where A_c, B_1, B_2 and B_3 are the peak amplitudes of the respective frequency components, ω_c is the carrier frequency and ω_1, ω_2 and ω_3 are the frequencies of the sinusoid components which make up the baseband or modulating signal.

Figure 6.1 shows an example of amplitude modulation. It can be observed that in accordance with the amplitude of the modulating signal the modulated signal magnitude varies between two curves. These are called the envelope of modulation and in fact show the shape of the modulating signal. Let us assume that the modulated output is given by $v(t)$. Then it follows from the principle of amplitude modulation that

$$v(t) = [A_c + m(t)] \cos \omega_c t$$

$$= A_c \cos \omega_c t + m(t) \cos \omega_c t$$

$$= A_c \cos \omega_c t + B_1 \cos \omega_1 t \cos \omega_c t + B_2 \cos \omega_2 t \cos \omega_c t$$
$$+ B_3 \cos \omega_3 t \cos \omega_c t \qquad (6.3)$$

Now using the trigonometric relations for the product of cosines, we have

$$v(t) = A_c \cos \omega_c t + \frac{B_1}{2} \cos (\omega_c + \omega_1) t + \frac{B_1}{2} \cos (\omega_c - \omega_1) t$$

$$+ \frac{B_2}{2} \cos (\omega_c + \omega_2) t + \frac{B_2}{2} \cos (\omega_c - \omega_2) t$$

$$+ \frac{B_3}{2} \cos (\omega_c + \omega_3) t + \frac{B_3}{2} \cos (\omega_c - \omega_3) t \qquad (6.4)$$

Fig. 6.2. Frequency spectra of modulating signal and amplitude modulated signal.

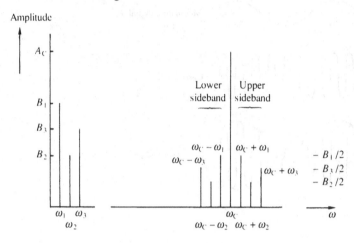

Equation (6.4) shows that the output consists of an unmodulated carrier and two groups of frequencies each of three components symmetrically positioned about the carrier frequency ω_c. The 'sum frequency' components constitute the upper sideband (USB) and the group containing the 'difference frequency' components is called the lower sideband (LSB). The bandwidth of the modulated signal is thus given by $[(\omega_c + \omega_3) - (\omega_c - \omega_3)]$ with a centre frequency at ω_c. It should be noted that information is contained in the sideband components only. Figure 6.2 shows the frequency spectra of the modulating signal $m(t)$ and the modulated signal $v(t)$.

Index of modulation

The index of modulation for a sinusoidal modulating signal is given by the ratio of the maximum deviation of amplitude from the unmodulated value to the amplitude of the unmodulated signal. For instance if the modulating signal is $B \cos \omega t$ and the carrier is $A_c \cos \omega_c t$ as shown in Fig. 6.3 then using the principle of the amplitude modulation as before the modulated output can be given by

$$v(t) = (A_c + B \cos \omega t)\cos \omega_c t = A_c(1 + B/A_c \cos \omega t)\cos \omega_c t \qquad (6.5)$$

From (6.5) we observe that the index of modulation is B/A_c. The percent modulation of an amplitude modulated wave is obtained by multiplying the index of modulation by 100. In practice modulation is maintained close to 100% thus allowing most of the baseband signal to be recovered by the

Fig. 6.3. Index of amplitude modulation.

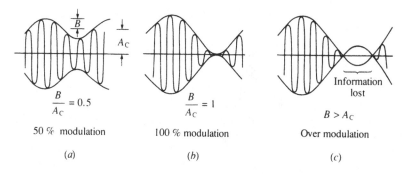

$\frac{B}{A_C} = 0.5$

50 % modulation

(a)

$\frac{B}{A_C} = 1$

100 % modulation

(b)

$B > A_C$

Over modulation

(c)

detector circuit at the receiving point. However, overmodulation, i.e. percent modulation greater than 100% switches off the carrier signal for part of the time as shown in Fig. 6.3(c) and hence distorts the modulating signal.

6.1.1 Modulation circuits

Modulation can be carried out at low power level or high power level. Both methods have their own advantages and disadvantages. In the former case relatively small modulating power is needed but the recovered information signal may be badly distorted. In high level modulation the baseband signal is amplified and the modulation takes place at the final stage. The method is efficient and there is little distortion in the recovered signal, but appreciable modulating power is necessary in order to achieve near 100% modulation.

Figure 6.4 shows a block diagram of low level amplitude modulation. A stable carrier frequency is provided by the oscillator which is isolated by a buffer so that any loading effect on the stability of the oscillator frequency is reduced. The intermediate power amplifier supplies the power gain needed

Fig. 6.4. Block diagram of low level amplitude modulation.

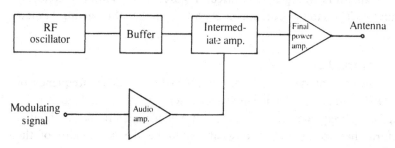

Fig. 6.5(*a*). A plate modulator for high power amplitude modulation.

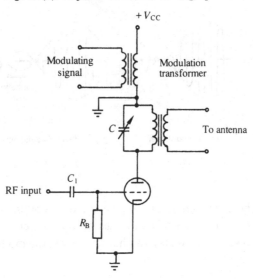

to drive the final power amplifier. In order to obtain highest efficiency, class C amplifiers are used in the final stage. Figure 6.5(*a*) shows a plate modulator for high power modulation. The circuit works as a class C amplifier. The modulating voltage $v_m(t)$ appears across the secondary of transformer T and thus is in series with the d.c. power supply V_{cc}. The effective power supply voltage v_{cc} is then the sum of $v_m(t)$ and V_{cc}. Let us consider that $v_m(t) = B\cos\omega t$, then

$$v_{cc} = V_{cc} + B\cos\omega t \tag{6.6}$$

The output will be proportional to the effective power supply and since the input is at carrier frequency we can write

$$v_0(t) \propto (V_{cc} + B\cos\omega t)\cos\omega_c t \tag{6.7}$$

where v_0 is the instantaneous output voltage of the modulator. Triode tubes are used for modulation in the kilowatt range. For an output of up to 100 watts, transistors are generally used. Figure 6.5(*b*) shows a collector modulator. The circuit works as a class C amplifier thus giving highest efficiency.

Worked example 6.1

An amplitude modulated waveform has a carrier frequency of 71.867 kHz and an upper sideband which extends from 72.2 kHz to 75 kHz. What is the frequency range of the lower sideband? If the modulated waveform has to be amplified what would be the bandwidth of the amplifier?

Fig. 6.5(*b*). A collector amplitude modulator.

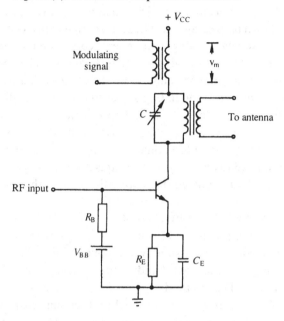

Solution

$$f_c + f_1 = 72\,200 \qquad f_c - f_1 = 71\,867 - 333 = 71.534\,\text{kHz}$$
$$f_c \qquad = 71\,867 \qquad f_c - f_3 = 71\,867 - 3133 = 68.734\,\text{kHz}$$
$$\overline{\quad f_1 = \quad 333\,\text{Hz}}$$

Therefore the lower sideband ranges
from 68.734 to 71.534 kHz.

$$f_c + f_3 = 75\,000$$
$$f_c \qquad = 71\,867$$
$$\overline{\quad f_3 = \quad 3133\,\text{Hz}}$$

Bandwidth of the amplifier:
$$75\,000 - 68\,734 = 6.266\,\text{kHz}.$$

6.1.2 Demodulation circuits

The modulated signal obtained from the antenna at the receiving
end is usually of the order of a few microvolts in magnitude and less than a
milliwatt in power. So the signal is first amplified with the help of a few
stages of tuned radiofrequency amplifiers as shown in Fig. 6.6. The diode *D*

Fig. 6.6. An AM demodulator circuit.

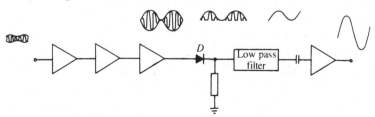

rectifies the amplified modulated signal. Assuming the diode to be ideal, only the positive halves of the modulated signal appear at the input of the low-pass filter as shown in the diagram. The rectified signal now contains a d.c. component, a slowly varying component corresponding to the modulating or baseband signal and also the carrier frequency. The low-pass filter has a cutoff frequency much smaller than the carrier frequency, but considerably higher than the modulating signal frequency. The diagram does not truly demonstrate the difference between the carrier and the modulating signal. In practice ω_c is at least several hundred times larger than ω_1, ω_2 and ω_3. The low-pass filter thus filters out the carrier frequency. Its output is then coupled to an amplifier by a capacitor which removes the d.c. components of the demodulated signal from the input to the amplifier.

6.1.3 Frequency division multiplexing

In communication systems very often several signals are combined together for simultaneous transmission over a single channel. This can be achieved by sharing an allocated bandwidth between a number of independent channels. The method is known as frequency division multiplexing (FDM). Figure 6.7(a) shows a FDM system where inputs from 12 channels are fed to a single outgoing channel. The information presented at various channels is amplitude modulated onto different carrier frequencies and the modulated signals are added together to form a composite signal which is then transmitted via a single channel. The separation between two adjacent carrier frequencies should be large

Fig. 6.7(a). Frequency division multiplexing.

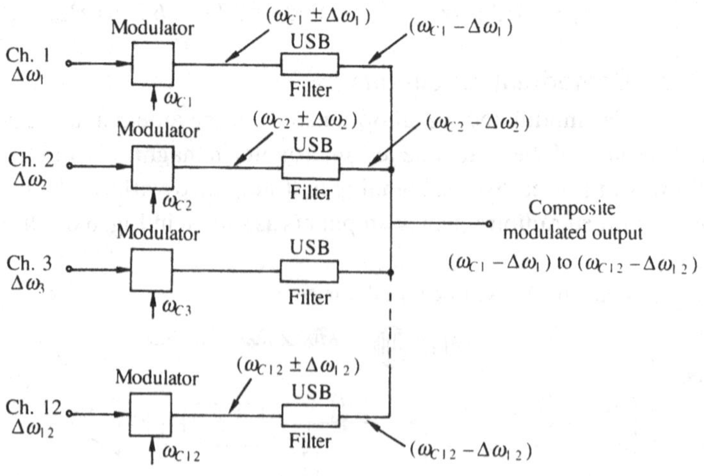

Fig. 6.7(*b*). Spectral occupancy of various amplitude modulated signals in the bandwidth (from ω_{c1} to ω_{c12}) after FDM.

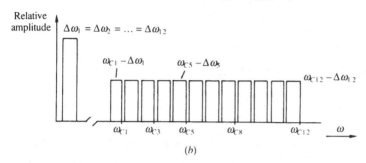

(*b*)

enough so that the upper sideband of one modulated signal does not overlap with the lower sideband of another modulated signal. Since the information can be retrieved from either of the upper and lower sidebands it is usual practice to filter out one sideband from each modulated signal as shown in the diagram, thus using less bandwidth per modulating signal. Figure 6.7(*b*) shows the spectral occupancy of various amplitude modulated signals in the bandwidth. Frequency division multiplexing is widely used for line communications (telephone systems) and also radio systems.

Worked example 6.2

Twenty speech signals concentrated in the band 300 Hz to 3.4 kHz are to be combined using frequency division multiplexing for transmission over a single communication channel. Estimate the bandwidth of the channel.

Solution

Allow a 4 kHz wide sideband for each speech signal. For double sideband transmission the channel should have:

$$20 \times 4\,\text{kHz} \times 2 = 160\,\text{kHz wide bandwidth}$$

For single sideband transmission the channel should have $20 \times 4\,\text{kHz} = 80\,\text{kHz}$ wide bandwidth.

6.2 Frequency modulation

In this method of modulation the frequency of the carrier wave changes in accordance with the amplitude of the modulating signal, but its amplitude remains constant, which is just the opposite of what happens in amplitude modulation. Although frequency modulation can be affected by relatively high frequency noise it is practically free from most types of low frequency noise which badly affects amplitude modulation.

Fig. 6.8. Frequency modulation.

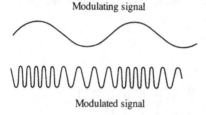

Figure 6.8 shows how a carrier wave is frequency modulated by an information signal. Let us suppose that the frequency of the carrier is 50 kHz, then at the instant when the amplitude of the modulating signal is zero, the modulated carrier frequency is 50 kHz and hence is called the centre frequency. With an increment of amplitude of the modulating signal the frequency of the carrier increases. Say the carrier frequency reaches 75 kHz when the modulating signal is at its positive peak. If the modulating signal is sinusoidal then at its negative peak the modulated signal frequency decreases to 25 kHz. Thus the frequency deviation from the centre frequency is 25 kHz for this particular modulating signal. The amount of deviation depends on the amplitude of the modulating signal.

The instantaneous frequency of a modulated signal is given by

$$\omega = \omega_c + k_f B \cos \omega_m t \tag{6.8}$$

where ω_c is the carrier frequency, k_f is a constant which determines the maximum frequency swing and $B \cos \omega_m t$ is the modulating signal. The total deviation of the modulated signal above and below the centre frequency is called the frequency swing. From Fig. 6.8 it can be observed that the bandwidth required by a communication channel to pass a frequency modulated signal is the sum of the maximum frequency swing and twice the frequency of the modulating signal. So we may write

$$\text{Bandwidth} = 2[(\omega - \omega_c)_{\text{max}} + \omega_m]$$

$$= 2[m_f + 1]\omega_m \tag{6.9}$$

where m_f, known as the index of modulation, is the ratio of the maximum frequency deviation $(\omega - \omega_c)_{\text{max}}$ produced by the modulating voltage B to the modulating frequency ω_m. From (6.8) and (6.9) we may write

$$m_f = (\omega - \omega_c)_{\text{max}}/\omega_m = k_f B/\omega_m$$

The above expression gives the relationship between the constant, k_f and the modulation factor, m_f.

Worked example 6.3

An FM signal has a frequency deviation of 30 kHz produced by a 3 V audio signal at 3 kHz. Calculate the modulation index and the required channel bandwidth.

Solution

$$m_f = \frac{\omega - \omega_c}{\omega_m}$$

$$= \frac{30 \times 10^3}{3 \times 10^3} = 10$$

$$\text{Bandwidth} = 2[m_f + 1]f_m$$

$$= 2[10 + 1] \times 3 \times 10^3$$

$$= 66 \, \text{kHz}$$

6.2.1 Modulation and demodulation circuits

A modified tuned collector oscillator can be used in order to achieve frequency modulation. The modulating signal is made to control the inductance or effective capacitance of the resonant circuit and thereby determine the instantaneous frequency. Figure 6.9 shows how a tuned circuit is formed with an L and C in conjunction with a varactor diode D_1. The latter connected in parallel with the capacitor C effectively control the frequency of oscillations. In the absence of any input signal to the circuit the oscillation frequency is the frequency of the transmitted carrier. However, when an audio (modulating) signal is applied to the input the capacitance of the varactor diode changes with its amplitude. The variation in the capacitance changes the frequency of oscillations thus in effect modulating the transmitted carrier.

The unmodulated resonant frequency f_0 of the LC circuit of Fig. 6.9 is given by

$$f_0 = 1/2\pi\sqrt{(LC)} \tag{6.10}$$

Fig. 6.9. A frequency modulation circuit.

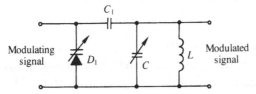

If the varactor capacitance has a maximum change of $\pm \Delta C$ then

$$f_{min} = 1/2\pi\sqrt{[L(C+\Delta C)]} \qquad (6.11)$$

and

$$f_{max} = 1/2\pi\sqrt{[L(C-\Delta C)]} \qquad (6.12)$$

From (6.10) and (6.11) we find the peak frequency deviation on the lower side of the centre frequency as follows

$$f_0 - f_{min} = f_0 - 1/2\pi\sqrt{[L(C+\Delta C)]} = f_0 - 1/2\pi\sqrt{[LC(1+\Delta C/C)]}$$

$$= f_0 - f_0/\sqrt{[(1+\Delta C/C)]} = f_0 - f_0(1+\Delta C/C)^{-1/2}$$

$$= f_0[1-(1+\Delta C/C)^{-1/2}] \simeq f_0[1-(1-\Delta C/2C)]$$

$$\simeq f_0 \Delta C/2C \qquad (6.13)$$

Similarly, the peak frequency deviation on the upper side of the centre frequency is,

$$f_{max} - f_0 = 1/2\pi\sqrt{[L(C-\Delta C)]} - f_0 = 1/2\pi\sqrt{[LC(1-\Delta C/C)]} - f_0$$

$$= f_0[(1-\Delta C/C)^{-1/2} - 1] \simeq f_0[1+\Delta C/2C - 1]$$

$$\simeq f_0 \Delta C/2C \qquad (6.14)$$

Therefore the peak frequency deviation is directly proportional to the change in capacitance of the varactor diode, which in turn changes in accordance with the amplitude of the modulating voltage. Now from the definition of the frequency swing we may write

$$f_{max} - f_{min} = (f_{max} - f_0) + (f_0 - f_{min})$$

Substituting from (6.13) and (6.14)

$$f_{max} - f_{min} = f_0 \Delta C/C \qquad (6.15)$$

Equation (6.15) enables us to find the bandwidth of the modulated signal as mentioned earlier in this section.

A simple demodulation circuit employing a phase-locked loop has already been described in Chapter 5 and shown in Fig. 5.16. In order to remove any amplitude variations on the FM signals caused by various low frequency noises, a limiting circuit may be used preceding the FM demodulation circuit.

6.3 Pulse modulation

In pulse modulation systems, a train of pulses is used as the carrier to transmit a modulating signal from one point to another. The mark–

space ratio of the pulses is considerably less than one, and this particular characteristic of the system enables a transmitter to deliver high values of peak power, the level of which can be several times larger than the average output power rating of the transmitter. There are several types of pulse modulation systems depending on the characteristics of the pulses, such as the amplitude, frequency, spacing or width, which are made to vary in accordance with the magnitude of the modulating signal.

Pulse amplitude modulation

Pulse amplitude modulation (PAM) is achieved by sampling the modulating signal by a train of pulses of constant amplitude as shown in Fig. 6.10. It can be noted that the signal is usually band limited with the help of sharp cutoff low-pass filters before being sampled by the train of pulses in order to avoid distortions (aliasing errors) in the recovered signals. Shannon sampling theorem states that a band limited signal $m(t)$ is fully described by sampling its amplitude at a rate equal to twice the highest frequency present in the signal. This can be easily explained with the aid of the diagrams in Fig. 6.11 which shows a sinusoidal signal the frequency of which we may assume to be the highest frequency appearing in a modulating signal. In Fig. 6.11(a) the sampling points are at intervals of greater than $1/2f_{max}$, and joining these points we obtain a trace which is not at all representative of the true curve. On the other hand in Fig. 6.11(b) the sampling points are at intervals of $1/2f_{max}$, and the line joining these points produces a trace which roughly describes the sinusoid.

Fig. 6.10. Pulse amplitude modulation.

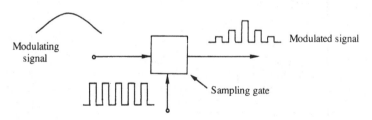

Fig. 6.11. Sampling points for (a) $f_s < 2f$ and (b) $f_s \geqslant 2f$.

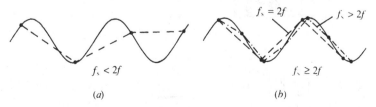

Therefore it can be said that it is quite unnecessary to transmit a modulating signal for all of the time, it can be present for only part of the time as a series of pulses. The original modulating signal can be reconstructed faithfully provided the sampling frequency is at least twice the highest frequency present in the modulating signal. A low-pass filter similar to one used prior to the sampling of the modulating signal can be used at the receiving end to reconstruct the signal.

Pulse frequency modulation

As the name implies pulse frequency modulation (PFM) is analogous to frequency modulation of a sinusoidal carrier. The repetition frequency of fixed-amplitude pulses is made to vary in accordance with the magnitude of the modulating signal. Figure 6.12(a) shows the form of pulse frequency modulation. The signal can be recovered using a similar technique to that used for frequency modulation of a sinusoidal carrier.

Pulse width modulation

In pulse width modulation (PWM) the width of the fixed-amplitude pulses is made to vary with the modulating signal as shown in Fig. 6.12(b). Low-pass filters can be used in this case in order to recover the information signal by taking the average of the modulated signal.

Fig. 6.12(a). Pulse frequency modulation. (b). Pulse width modulation. (c). Pulse position modulation.

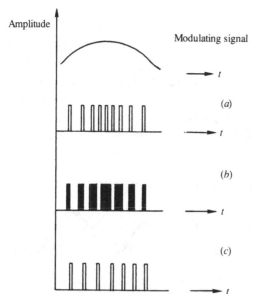

Pulse position modulation

In pulse position modulation (PPM) the pulses are of constant width and amplitude. Each of the pulses is advanced or delayed relative to a point within a time slot according to the amplitude of the modulating signal. Figure 6.12(*c*) shows the form of this type of modulation.

6.4 Pulse coded modulation

In the last section all the different types of pulse modulation we have discussed are analogue types, whereas pulse coded modulation is of a digital form. At regular intervals the amplitude of the modulating signal is sampled and then converted to a digital code made up of binary digits 0 and 1. Usually groups of four binary digits represent the amplitudes of a signal. Thus 2^4 or 16 levels of amplitude can be transmitted by a serial string of binary digits. In a group the absence of a pulse denotes binary digit 0, whereas the presence of a pulse represents binary digit 1. Figure 6.13 shows an example of pulse coded modulation. Since the train of pulses has only two distinct levels of magnitude, i.e. '1' and '0', this system can provide virtually error-free transmission in very noisy channels. In order to change between analogue and digital signal, A-to-D and D-to-A converters are used. Pulse coded modulation is widely used in transcontinental telephone systems, where the signal is amplified many times along the way. Each time before amplification, binary codes are reconstructed correctly, thus avoiding any possibility of generating errors.

6.5 Time division multiplexing

In frequency division multiplexing, discussed earlier in this chapter, an allocated bandwidth is shared by a number of independent signals at all time. On the other hand, in time division multiplexing the total bandwidth is available to all signals, but the time is shared by them.

Fig. 6.13. Pulse coded modulation.

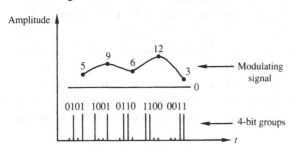

Fig. 6.14. Time division multiplexing.

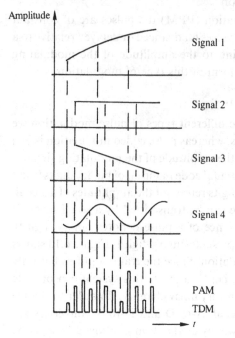

Figure 6.14 shows how a composite time division multiplexed signal for a single channel can be obtained from four independent modulating signals using an analogue multiplexer which can connect several input channels sequentially onto a single channel with the help of a control circuit. The principles of multiplexers will be discussed in detail in the next chapter.

Summary

1. Low frequency information signals are carried from one place to another by radio frequency waves. Various types of modulation are used to transpose low frequency information signals onto the carrier waves.

2. In amplitude modulation the amplitude of the information signal, also called the modulating signal, determines the instantaneous amplitude of the radio frequency carrier waves. The modulating frequency determines the rate of amplitude changes in the carrier waves.

3. An amplitude modulated carrier has two groups of frequencies symmetrically positioned about the carrier frequency. They are the sums and differences of the carrier and modulating frequencies and are known as

the upper sideband (USB) and lower sideband (LSB) respectively. Information signals can be recovered at the receiving end from either of these sidebands.

4. Index of modulation for amplitude modulation is defined as the ratio of the maximum deviation of amplitude from the unmodulated value to the amplitude of the unmodulated signal. For good quality transmission it should be maintained close to one. However it should not be larger than one otherwise overmodulation takes place thus giving rise to partial loss of information.

5. The instantaneous output voltage of an amplitude modulator is proportional to $(V_{cc} + B \cos \omega t) \cos \omega_c t$, where V_{cc} is the d.c. power supply, B and ω are the amplitude and frequency of the modulating signal respectively and ω_c is the carrier frequency.

6. A low-pass filter in conjunction with a diode acts as a simple demodulator for an amplitude modulated wave.

7. Frequency division multiplexing allows several signals to share an allocated bandwidth and to be transmitted all at the same time.

8. In frequency modulation the amplitude of the modulating signal determines the instantaneous carrier frequency, whereas the frequency of the modulating signal determines the rate of frequency deviations in the carrier.

9. In frequency modulation, the index of modulation is the ratio of the maximum frequency deviation produced by the modulating voltage to the modulating frequency.

10. A tuned collector oscillator can perform frequency modulation when a varactor diode is added to the LC tuning circuit.

11. In pulse modulation systems, a train of pulses is used as the carrier to transmit an information signal from one place to another. The pulse modulation enables a transmitter to deliver high values of peak power which may be several times larger than the average output power rating of the transmitter.

12. There are several types of pulse modulation systems depending on the characteristics of the pulses such as the amplitude, frequency, spacing and width which are made to vary in sympathy with the amplitude of the information signal.

13. Pulse coded modulation uses digital codes made up of binary digits 0 and 1. It provides error-free transmission in noisy communication channels and is extensively used in transcontinental telephone systems.

14. In time division multiplexing the total allocated bandwidth is available to all signals, but the time is shared by them.

Problems

6.1. An amplitude modulator is driven by an 810 kHz carrier and has an audio signal input containing frequency components between 150 Hz and 10 kHz. What frequency components are in the lower sideband of the AM output and also in the upper sideband?

6.2. An amplitude modulated waveform has a carrier frequency of 290 kHz and an upper sideband which extends from 291.2 kHz to 292.5 kHz. What is the frequency range of the lower sideband? What range of frequencies must be included in the passband of an amplifier that will be used to amplify the modulated signal?

6.3. A 10 volt RMS, 230 kHz carrier drives an amplitude modulator and is modulated by a 2.5 volt RMS, 6 kHz sinusoidal signal. What is the expression for the modulated output?

6.4. A signal $0.9 \sin 2\pi 10^3 t$ modulates a carrier of $10 \sin 2\pi 310 \times 10^3 t$ in amplitude. What is the index of modulation as a percentage? Suggest a way by which the index of modulation can be increased to 100%.

6.5. Eight speech signals concentrated in the band 250 Hz to 3 kHz, and ten 50 Hz distorted signals having up to 9th harmonic components are to be combined using frequency division multiplexing for transmission over a single communication channel. Estimate the bandwidth of the channel for (a) single sideband and (b) double sideband transmission.

6.6. A communication channel with a band ranging from 20 kHz to 65 kHz is available to transmit audio signals from one point to another. There are six signals which have to be transmitted and their frequency ranges are as follows: 300–3400 Hz, 300–7500 Hz, 300–3400 Hz, 300–5000 Hz, 300–10 000 Hz and 300–3400 Hz. Design a suitable frequency division multiplex system.

6.7. Show, how a tuned collector oscillator, can perform frequency modulation with the help of a varactor diode.

6.8. A frequency modulated signal has a frequency deviation of 25 kHz produced by a 2 volt peak audio signal at 3.5 kHz. Calculate the modulation index and the required channel bandwidth.

6.9. A modulating signal has an amplitude of 3 volt peak and frequency of 2 kHz. The carrier wave has an amplitude of 4 volt peak and frequency of 50 kHz. If the frequency modulation is used for transmission of the signal write an expression for the modulated wave. The frequency deviation produced by the modulating signal is 20 kHz.

Fig. P.6.10

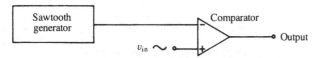

6.10. The pulse-width modulator in Fig. P.6.10 is driven by 150 Hz sawtooth. When v_{in} is $+2$ volts the modulator output is a series of pulses having widths equal to 1.5 μs. What are the pulse widths when v_{in} is $+4.5$ volts?

7

Data acquisition and distribution

Objectives

At the end of the study of this chapter a student should:

1. be able to design different types of digital-to-analogue (D-to-A) converters
2. be familiar with various kinds of analogue-to-digital (A-to-D) converters; their advantages and disadvantages and be able to construct them
3. be familiar with the errors in the converters
4. be conversant with the design of discrete multiplexers and demultiplexers and be familiar with their limitations and errors
5. be familiar with the principles of sample-and-hold circuit and be able to design them
6. be able to develop a suitable data acquisition or distribution system for a particular need and find the accuracy of such a system.

In order to process analogue signals which are continuous and of varying magnitude over a period of time, with the aid of computers, conversion of analogue currents or voltages into digital codes is essential. Again, conversely, in order to control machines with the help of computers, digital signals have to be converted into analogue currents and voltages. A typical system is shown in Fig. 7.1. Here the analogue sensors measure physical quantities such as temperatures, pressures, etc., and the analogue controls switch on (or off) heaters, pumps and so on.

Sometimes it is necessary to acquire or distribute more than one signal simultaneously. This can be achieved by using the time-sharing techniques. The time required for acquiring data from many sources, or distributing data to many controls can be reduced dramatically if a single channel is time-shared for transferring data. In this chapter different types of circuits, which are necessary to build a data acquisition or distribution system will be discussed.

Fig. 7.1. A typical data acquisition and distribution system.

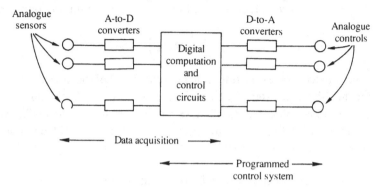

7.1   Digital-to-analogue (D-to-A) converters

A D-to-A converter is in fact a decoder. It converts a digital (coded) signal X_d into an analogue signal V_x according to the expression

$$V_x = V_R X_d \tag{7.1}$$

where V_R is an analogue reference voltage. This is the basic transfer function for a D-to-A converter. Figure 7.2 shows the transfer characteristic of a 4-bit binary D-to-A converter. For such a converter, (7.1) can be rewritten in the form,

$$V_x = V_R \left[\frac{1}{2}a_1 + \frac{1}{2^2}a_2 + \frac{1}{2^3}a_3 + \frac{1}{2^4}a_4 \right] \tag{7.2}$$

Fig. 7.2. Transfer characteristic of a 4-bit D-to-A converter.

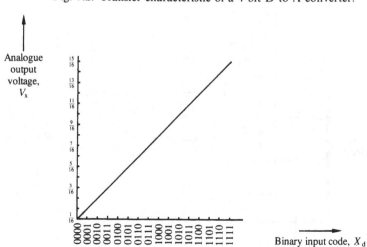

where a_1, a_2, a_3 and a_4 represent the binary number '0' or '1'. For instance, if $a_1 = a_2 = a_3 = a_4 = 1$, then we will have the maximum analogue voltage,

$$V_{x(max)} = V_R \left(\frac{1}{2} + \frac{1}{2^2} + \frac{1}{2^3} + \frac{1}{2^4} \right) = V_R (1 - 1/2^4) \simeq V_R \qquad (7.3)$$

Thus V_R determines the full-scale reading (FSR) of the device. In this case, for the 4-bit converter, the least significant bit (LSB) in the binary code is $1/2^4$. For an n-bit converter the LSB will be $1/2^n$. By observing (7.3) we may say that the analogue voltage which is equivalent to one LSB in the binary code is given by $1/2^n \cdot$ FSR.

7.1.1. Weighted-resistor D-to-A converter

This is the simplest and most straightforward of all D-to-A converters. Figure 7.3 shows the circuit of an 8-bit converter which has the same number of electronic switches $(S_1 - S_8)$ as the number of bits. The resistors connected to the switches have values in an increasing binary weighted sequence. Each of eight electronic switches is controlled by one bit line. When the bit corresponding to a switch is a '1', then the switch connects the reference voltage $- V_R$ to the respective resistor, the other end of which forms the summing point of an operational amplifier, and a current flows through the resistor. If the bit is a '0' then the switch connects the resistor to the ground and hence no current flows through that particular resistor to the summing point. If all the switches are on, currents with magnitudes of

Fig. 7.3. An 8-bit weighted-resistor D-to-A converter.

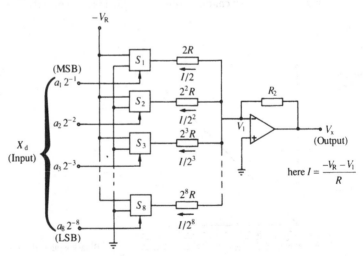

$$\frac{-V_R-(V_1)}{2R}, \; \frac{-V_R-(V_1)}{2^2R}, \; \frac{-V_R-(V_1)}{2^3R}, \cdots \frac{-V_R-(V_1)}{2^8R}$$

are generated. The operational amplifier has high input impedance and very large gain, therefore, the two usual assumptions, i.e., no current flow into the input terminals of the operational amplifier and zero differential-mode voltage, can be used to obtain the analogue voltage V_x as follows,

$$V_x = V_R/R\left(\frac{1}{2} + \frac{1}{2^2} + \frac{1}{2^3} + \cdots \frac{1}{2^8}\right)R_2 \tag{7.4}$$

where R_2 is the feedback resistor of the amplifier.

Since the operational amplifier is being connected in the inverting mode, a negative reference voltage $-V_R$ has been used in order to obtain a positive analogue voltage at the output. By adjusting the factor $V_R/R \cdot R_2$ we can adjust the FSR of the device. The resistor corresponding to the most significant bit (MSB) has a value $2R$ and that corresponding to the LSB has a value 2^8R.

The overall accuracy of this converter depends on (i) the input offset current and voltage of the operational amplifier, (ii) the tolerance of the reference voltage supply, (iii) the tolerance of ratios of the resistor values and also (iv) the error due to the 'on' resistance r_{on} of the FET used in the electronic switches.

The speed of conversion depends on the slew rate of the operational amplifier and also the turn-on and turn-off times of the switches. The main drawback of this type of converter is that it needs a very wide range of resistor values. For an 8-bit converter the ratio of the largest to the smallest resistor would be $2^8R/2R = 2^7 = 128$. In discrete circuits this may not cause any problems, but in thin-film, or monolithic integrated circuits where fabrication is carried out with material of one resistivity, it is difficult to manufacture such a wide range of resistor. Therefore, this type of D-to-A converter is used only for a relatively small number of bits. If the number of bits is large, then a different type of D-to-A converter which will be discussed in the next section should be used.

The electronic switches of the converter can be operated by connecting two N type MOSFETs as shown in Fig. 7.4. When the digital input is a '1', the FET T_1 is on and the FET T_2 is off since the NOT gate inverts the signal to T_2. Thus the reference voltage, V_R is applied to the resistor, 2^nR. On the other hand, if the binary input is a '0', then T_1 is off and T_2 is on. Thus the resistor, 2^nR is grounded. The FETs have finite 'on' resistances which give rise to errors; however, the value of $2R$ is, in practice, much larger than r_{on} and thus makes the error negligible.

Fig. 7.4. Electronic switches for the converters.

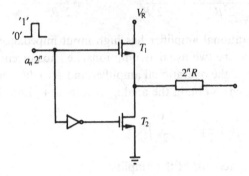

Worked example 7.1

The reference voltage for a 6-bit weighted-resistor D-to-A converter is obtained by using the breakdown voltage of a Zener diode. If the FSR of the converter is 12 volts, the smallest weighted resistor has a value of 20 kΩ and the feedback resistance is 40 kΩ, what is the breakdown voltage of the Zener diode? If the converter has a binary input 101101, what would be the output?

Solution

Assuming all the switches are on,

$$V_R\left(\frac{1}{2\times10\times10^3}+\frac{1}{2^2\times10\times10^3}+\cdots\frac{1}{2^6\times10\times10^3}\right)\times40\times10^3=12$$

or

$$V_R\left(1-\frac{1}{2^6}\right)\frac{40\times10^3}{10\times10^3}=12$$

$$\therefore\qquad V_R=V_{\text{Zener}}\simeq\tfrac{12}{4}=3 \text{ volts}$$

$$V_x=V_R\left(\frac{1}{2\times10\times10^3}+\frac{1}{2^3\times10\times10^3}+\frac{1}{2^4\times10\times10^3}\right.$$

$$\left.+\frac{1}{2^6\times10\times10^3}\right)40\times10^3$$

$$=3(\tfrac{1}{2}+\tfrac{1}{8}+\tfrac{1}{16}+\tfrac{1}{64})\times4=8.4375 \text{ volts}$$

Worked example 7.2

In the circuit of Fig. 7.3 the value of the smallest weighted resistor is 25 kΩ. What should be the maximum allowable FET 'on' resistance for an error of only 1%? Neglect the other factors which may also affect the accuracy.

Solution

$$\left(\frac{R + r_{\text{on}}}{R}\right) \times 100 = 100 + 1$$

or,

$$R + r_{\text{on}} = 1.01R$$

$$\therefore \qquad r_{\text{on}} = 0.01 \times 25 \times 10^3 = 250\,\Omega$$

7.1.2. *R–2R* ladder D-to-A converter

In order to understand the operation of this type of converter, it is essential to be familiar with the basic principle of the resistor-ladder network as shown in Fig. 7.5. The main characteristic of this network is that the resistance seen by any of the voltage sources is equal to $3R$. Thus the current flowing out from any source is equal to $V/3R$. By close examination it can also be found that a current at each node divides equally into two components. Current I_4 becomes $I_4/2$ on either side of node N_4 and subsequently becomes $I_4/2^2$ after node N_3 and so on. Therefore, at each node the current decreases in a binary weighted sequence. The total current in the load resistor R_L is as shown in the figure.

Fig. 7.5. Resistor-ladder network.

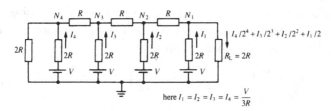

A D-to-A converter circuit which uses this property of the R–$2R$ ladder network is shown in Fig. 7.6. The circuit consists of eight electronic switches which are controlled by the binary input signal in a similar manner to that in the weighted-resistor type. However, in contrast to the weighted-resistor D-to-A converter, the precision resistors in the R–$2R$ ladder D-to-A converter have only two values, R and $2R$. In practice, a value of $20\,\text{k}\Omega$ is chosen for R in thin-film fabrication, the reason mainly being the size and the availability of the material.

Fig. 7.6. R–$2R$ ladder D-to-A converter.

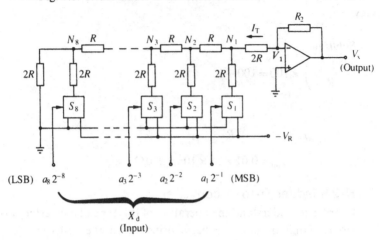

(LSB) $a_8 2^{-8}$ $a_3 2^{-3}$ $a_2 2^{-2}$ $a_1 2^{-1}$ (MSB)

X_d
(Input)

When a binary bit is a '1', the corresponding switch is on and the reference voltage is connected to the respective resistor, thus providing a current $I_n = -V_R/3R$ through it. This current is divided equally into two components at node 'N_n'. The component which flows towards the summing point of the operational amplifier becomes divided equally again at node N_{n-1}. For instance, when the input terminal $a_8 2^{-8}$ has a '1' the current $I_8 = V_R/3R$ is split into two equal components eight times, once at each node, by the time it reaches the summing point of the operational amplifier. Similarly, the current due to a '1' at the binary input terminal $a_7 2^{-7}$ is split into two equal components seven times. By using the superposition theorem we can express the total current flowing into the summing point of the operational amplifier,

$$I_T = \frac{(V_R - V_1)}{3R}\left[\frac{1}{2} + \frac{1}{2^2} + \frac{1}{2^3} + \cdots + \frac{1}{2^8}\right] \tag{7.5}$$

Since an operational amplifier has a very high input impedance, all of the current passes through the feedback resistor R_2 thus yielding an analogue voltage at the output of the operational amplifier,

$$V_x = R_2 I_T = \frac{(V_R - V_1)}{3R}\left[\frac{1}{2} + \frac{1}{2^2} + \cdots + \frac{1}{2^8}\right]R_2 \tag{7.6}$$

Again an operational amplifier has a very high gain and we may assume, $V_x \gg V_1$; then (7.6) can be rewritten as follows

$$V_x = \frac{V_R \cdot R_2}{3R}\left[\frac{1}{2} + \frac{1}{2^2} + \cdots + \frac{1}{2^8}\right] \tag{7.7}$$

Fig. 7.7. Inverted ladder D-to-A converter.

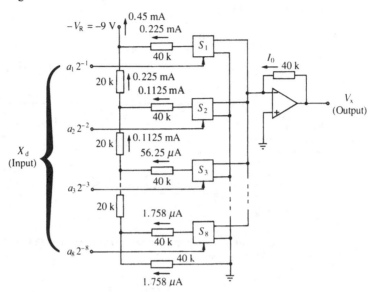

The FSR can be easily changed by varying the feedback resistor R_2. The scale factor error can also be reduced by varying R_2 as will be discussed later in this chapter. Errors due to the input offset voltage and current of the operational amplifier can be minimised by using standard operational amplifier offset correction methods as discussed in Chapter 3. The main advantage of this type of converter is that the ratio of the largest to the smallest resistor required is only 2 which can be easily implemented by the thin-film or monolithic integrated-circuit technology.

An inverted ladder D-to-A converter is shown in Fig. 7.7. Here the reference voltage is connected to what is normally the output of a ladder network, and the terminals used normally as the inputs are grounded. By examining Fig. 7.7 we can see that this arrangement causes currents to flow in a binary weighted sequence through the $2R$ resistors. Switches select the appropriate binary fractions of currents according to a binary input code, and these currents are then added together at the summing point of an operational amplifier. Due to the high input impedance of the operational amplifier, all of the total current flows through the feedback resistor thus developing an analogue voltage equivalent to the binary input code. In the circuit of Fig. 7.7 the reference voltage is $-9\,\mathrm{V}$ and the full-scale output current is given by

$$I_0 = V_R/R[1/2 + 1/2^2 + \dots + 1/2^8]$$

$$= V_R/R[1 - 1/2^8] = 9/20 \times 10^3 \times 255/256\,\mathrm{A}$$

Fig. 7.8. Block diagram of RS DAC0800 type D-to-A converter. (RS Components Ltd.)

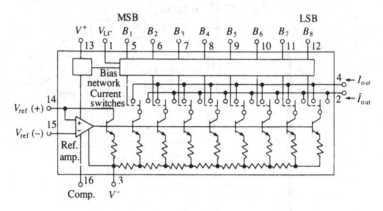

Fig. 7.9. Low input impedance output circuit for RS DAC0800 type converter. (RS Components Ltd.)

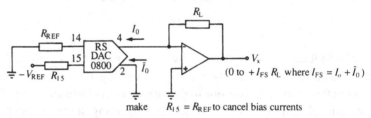

The RS DAC0800 type D-to-A converter works on this principle. It is an 8-bit high speed current output D-to-A converter. The digital inputs are directly TTL compatible but simple circuitry enables the device to interface with most common logic families. Figure 7.8 shows the block diagram of the RS DAC0800. By adding an operational amplifier to its output as shown in Fig. 7.9, a low impedance analogue output can be obtained. The relationship between the output currents and related circuit variables for the positive output operation is as follows,

$$I_0 + \bar{I}_0 = I_{FS} \tag{7.8}$$

$$I_{FS} \simeq - V_{REF}/R_{REF} \times 255/256 \, \text{mA} \tag{7.9}$$

where

R_{REF} is the reference resistance in kΩ
V_{REF} is the reference voltage in volts

and

I_{FS} is the full-scale current

The output voltage swing can be adjusted by varying R_L to suit individual requirements.

The D-to-A converter output takes a finite time, known as its settling time, to reach a final output value and settle out within a specified band of error whenever there is a change in the binary input code. The specified band of errors is usually $\pm\frac{1}{2}$LSB. The worst case settling time is the duration during which the output of a D-to-A converter changes from zero to maximum value. The RS DAC0800 has a typical settling time of 100 ns.

Large switching transients, commonly known as glitches, occur at the output even when the binary input code changes by only 1 bit, e.g. from 1000 to 0111. It happens, due to the fact that the switches' turn on and turn off times are always unequal. For this particular example, if the MSB changes from '1' to '0' before the other bits change to '1's from '0's, then the output of the D-to-A converter will momentarily drop to zero (0000) before becoming equivalent to 0111 which is nearly the half of the FSR, thus producing a very large transient.

Worked example 7.3

The output of the circuit of Fig. 7.6 is connected to an analogue meter having a FSR of 10 volts. What is the maximum value the 'R_2' can have if the resistors used for the ladder network are of 15 kΩ and 30 kΩ values and the reference voltage is -9 V?

Solution
Maximum V_x is 10 V. From (7.7)

$$V_x = \frac{V_R \cdot R_2}{3R}\left[\frac{1}{2} + \frac{1}{2^2} + \cdots + \frac{1}{2^8}\right]$$

Fig. ex.7.3

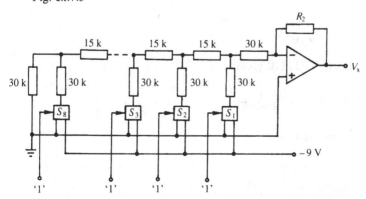

or

$$10 = \frac{9 \cdot R_2}{3 \cdot 15 \cdot 10^3} \left[\frac{1}{2} + \frac{1}{2^2} + \cdots + \frac{1}{2^8} \right]$$

Therefore, the maximum value of R_2 is

$$\frac{45 \times 10^4}{9} \times \frac{256}{255} = 50.2 \, k\Omega$$

Worked example 7.4

The digital input 11011011 applied to a RS DAC0800 causes it to produce an output current $I_0 = 4.275 \, mA$. Find the value of the D-to-A converter reference current and the value of the complementary output current $\bar{I}_0$. Determine also the values of I_0 and $\bar{I}_0$ for the digital inputs 10000001, and 01111110.

Solution

$$I_0 = I_{REF} \left[\frac{1}{2} + \frac{1}{2^2} + \frac{1}{2^4} + \frac{1}{2^5} + \frac{1}{2^7} + \frac{1}{2^8} \right] = \frac{219}{256} I_{REF}$$

Given $I_0 = 4.275 \, mA$.

$$\therefore \qquad I_{REF} = \frac{4.275 \times 256}{219} = 5 \, mA$$

$$\bar{I}_0 = \tfrac{36}{256} I_{REF} = 0.703 \, mA$$

Digital input 10000001 gives:

$$I_0 = 5 \times 10^{-3} \left[\frac{1}{2} + \frac{1}{2^8} \right] = 2.52 \, mA$$

and

$$\bar{I}_0 = 5 \times 10^{-3} \times \tfrac{126}{256} = 2.46 \, mA$$

Digital input 01111110 gives:

$$I_0 = 5 \times 10^{-3} \left[\frac{1}{2^2} + \frac{1}{2^3} + \frac{1}{2^4} + \frac{1}{2^5} + \frac{1}{2^6} + \frac{1}{2^7} \right]$$

$$= 2.46 \, mA$$

and

$$\bar{I}_0 = 5 \times 10^{-3} \times \tfrac{129}{256} = 2.52 \, mA$$

Fig. 7.10. Transfer characteristic of a 4-bit A-to-D converter.

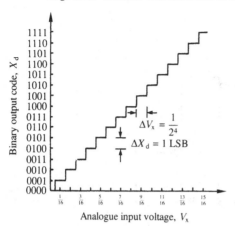

Analogue input voltage, V_x

7.2 Analogue-to-digital (A-to-D) converter

An A-to-D converter is in essence an encoder. It converts an analogue signal V_x into a binary code X_d. They are related by the expression

$$X_d = \frac{V_x}{V_R} \qquad (7.10)$$

where V_R is an analogue reference voltage and X_d represents a fractional binary number $X_d = a_1 \frac{1}{2} + a_2 \frac{1}{2^2} + \dots + a_n \frac{1}{2^n}$

The transfer characteristic of a 4-bit A-to-D converter is shown in Fig. 7.10 from which we may note that the analogue signal can be approximated with the digital output X_d only to the smallest increment ΔV_x where $\Delta V_x = 1/2^n \cdot V_R$ since $\Delta X_d = 1/2^n$.

There are a great many circuits that can perform the A-to-D conversion. In this section we shall discuss only a few types, some of which are very simple to design and construct but not so accurate, whereas the others are of complex nature but meet the demands of high accuracy.

7.2.1 Parallel A-to-D converter

This type of A-to-D converter is the simplest of all as can be seen from Fig. 7.11. In this circuit the analogue voltage is fed simultaneously to one input of each of several comparators, the number of which depends on the desired number of bits in the binary output code.

The other inputs of the comparators are connected to equally spaced reference voltages which may be obtained by connecting a series of resistors

Fig. 7.11. Parallel A-to-D converter.

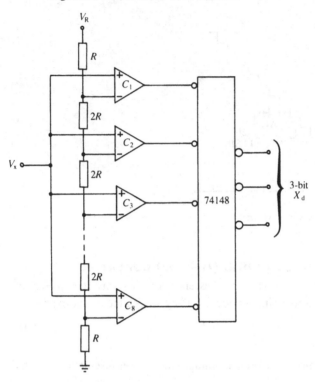

to a reference voltage V_R as shown in the figure. For this particular circuit the reference voltages applied to the comparators starting from the top are $15V_R/16, 13V_R/16, 11V_R/16 \ldots V_R/16$. The output of a comparator will be a '1' when the analogue input voltage is greater than the reference voltage applied to that comparator, otherwise it will be a '0'. The outputs from the eight comparators are fed to a logic circuit in order to obtain the desired binary coded output in three bits. A 74148 type encoder may be used as the logic circuit as shown in the diagram. Very high conversion speeds can be achieved with this type of converter by choosing extremely fast comparators.

Worked example 7.5

What would be the resolution of the parallel A-to-D converter shown in Fig. 7.11 if the number of comparators in the circuit is 16 and the reference voltage is 10 volts?

Solution

The difference between reference voltages applied to any two adjacent comparators is

$$\frac{V_R}{32R} \cdot 2R = \frac{V_R}{16}$$

Therefore, the resolution in volts is $\frac{10}{16} = 0.625$ volts.

7.2.2. Servo A-to-D converter

The servo A-to-D converter is the simpler of the two very popular parallel feedback A-to-D converters which use D-to-A converters in their feedback paths. It has a moderate speed of conversion and is a compromise between the two for continuously changing analogue signals. The device has three basic circuits: (i) a comparator, (b) an up and down counter and (c) a D-to-A converter, as shown in Fig. 7.12. The comparator compares V_x, the analogue input voltage with V_f, the analogue output voltage of the D-to-A converter. $R/2R$ ladder type D-to-A converters are commonly used by the manufacturers in the feedback path. The input of the D-to-A converter is, in fact, the binary coded output, X_d of the whole device. The output of the comparator is connected to the up and down counter as shown in the diagram. Suppose at an instant, V_x is larger than V_f, i.e., the binary coded output represents an analogue voltage smaller than the input analogue voltage V_x, then the output of the comparator produces a '1'. This makes the counter count up. It counts up until X_d correctly represents the analogue input voltage V_x. Then, of course, V_f will be equal to V_x within $1/2^n$ of the analogue FSR of the device which is equivalent to the LSB of the binary code.

If V_x now decreases, V_f becomes greater than V_x and the comparator

Fig. 7.12. Servo A-to-D converter.

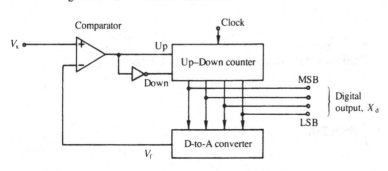

output produces a '0'. It is converted to a '1' by the inverter in order to make the counter count down until X_d again becomes equivalent to V_x.

The conversion time of the converter is directly proportional to the input voltage change. Only one clock pulse is required to record a change in analogue input voltage equivalent to one LSB, whereas, 2^{n-1} clock pulses would be needed to record a full-scale change.

The output of the D-to-A converter can be incremented or decremented by only a LSB during a single clock period. Therefore, its input should not change by more than an amount equivalent to the LSB in one clock period. Otherwise, errors will occur at the output of the D-to-A converter and hence the output of the A-to-D converter. In other words we may say:

maximum rate of change in full scale input signal $< 1/2^n$ FSR f_c

$\mathrm{d}/\mathrm{d}t$ (FSR/2 $\sin 2\pi ft)_{\max} < 1/2^n$ FSR f_c

$$f_{\max} < \frac{f_c}{2^n \pi} \qquad\qquad (7.11)$$

A finite time is required by the comparator to change state due to the slew of the output of the operational amplifier used in the comparator, and thus the slew rate of the operational amplifier imposes a limit on the clock frequency applied to the counter. With high speed operational amplifiers, clock frequency as high as several MHz can be used, but then the propagation delay of the cascaded AND gates in the counter, and the r_{on} of the switches and the parasitic reactances of the ladder network become the limiting factors on the speed.

Worked example 7.6

The FSR of a 6-bit servo type A-to-D converter is 10 volts. A sinusoidal signal of ± 2 volts amplitude is applied to its input. What is the maximum frequency the input signal can have if the clock frequency is 1 MHz?

Solution

Maximum rate of change in the input signal $= \dfrac{\text{FSR}}{2^n} f_c$,

or

$$\frac{\mathrm{d}}{\mathrm{d}t} |2 \sin 2\pi ft|_{\max} = \frac{10}{2^6} \times 10^6$$

or

$$2 \cdot 2\pi f_{max} = \frac{10^7}{2^6}$$

$$\therefore \quad f_{max} = \frac{10^7}{2 \times 2\pi \times 2^6} = 12\,440\,\text{Hz}$$

7.2.3 Successive approximation A-to-D converter

The successive approximation A-to-D converter is also a parallel feedback type A-to-D converter. Nearly all the A-to-D converters manufactured nowadays use the successive approximation technique. It is a compromise between cost and accuracy. Here again the output binary code is converted back to an equivalent analogue voltage, with the aid of a D-to-A converter, to be compared with the input analogue voltage. The binary code is made approximately equivalent to the input analogue voltage in a sequence of successive steps the number of which depends on the number of bits of the device. The method of conversion by this converter is similar to the process of weighing on an ordinary chemist's balance where in order to find the weight of a substance all the reference weights are tried on the balance one by one. First the heaviest weight is tried. If the substance weighs more than this then the heaviest weight is left on the balance and the next heaviest weight which is usually one half of the heaviest weight is put on the balance. It is left on the balance or taken off depending on whether the substance is heavier or lighter than these two weights added together. Then the next heaviest weight, i.e., one quarter of the heaviest is tried. The process is continued until all the available reference weights have been tried. The sum of the reference weights left on the balance then represents the closest approximate weight of the substance.

Similarly in a successive approximation A-to-D converter the most significant bit is tried first during the first clock period. It is left in or taken out from the storage circuit depending on whether the input voltage is greater than the analogue equivalent of the output binary code or not. Then during the next clock period the next most significant bit is tried. The process is repeated until all the bits have been dealt with.

Let us now consider that an analogue voltage of 7.12 volts has been applied to a 4-bit converter. Suppose the R–$2R$ ladder D-to-A converter's reference voltage V_R is -10 volts, then the feedback voltage from the D-to-A converter for each of the four bits starting from the MSB is equivalent to 5, 2.5, 1.25 and 0.625 volts respectively. At first we try the most significant bit. This means that the MSB is set to '1' and all the other three bits are set

Fig. 7.13. Successive approximation A-to-D converter.

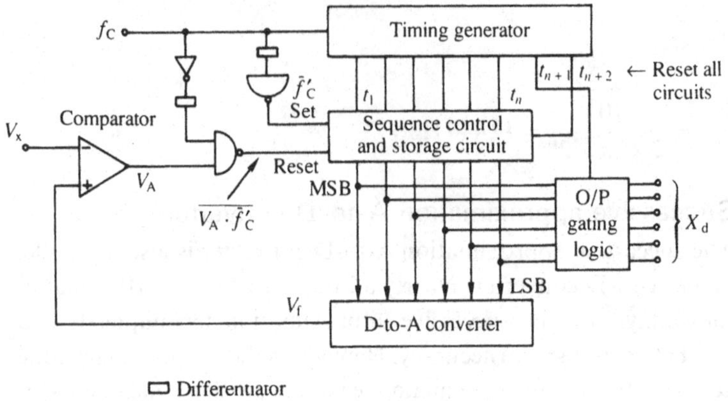

☐ Differentiator

to '0' which is equivalent to an analogue voltage of $(5+0+0+0)=5$ V. The input voltage is greater than this, so the MSB remains at '1' and the next MSB is made '1', i.e., we try 1100 during the next clock period. Now the feedback voltage is $(5+2.5)=7.5$ volts which is greater than the input voltage. So this bit is made '0' again (analogous to a reference weight taken off the chemist's balance). Now we make the next MSB equal to '1' and the binary output consequently becomes 1010. This gives us a feedback voltage of the magnitude $(5+0+1.25+0)=6.25$ volts. This is less than the input voltage and hence the third bit remains at '1'. We try now with the LSB equal to '1'. The feedback voltage is now $(5+0+1.25+0.625)=6.875$ volts which is smaller than the input voltage and the LSB remains at '1'. Since we have no more bits to try the output binary code is 1011. It should be noted that the difference in the analogue input voltage and the analogue equivalent of the binary coded output is $7.12-6.875=0.245$ volts. This difference, i.e., the quantisation error which will be discussed later in this chapter can be reduced by using a converter with a larger number of bits.

A typical successive approximation circuit consists of a comparator, a timing generator, a sequence control and storage circuit, a D-to-A converter and an output gating logic as shown in Fig. 7.13. The clock pulses control and synchronise the operation of the circuit. The comparator compares the feedback signal V_f from the D-to-A converter with the input analogue voltage V_x.

When the analogue voltage V_x is greater than the feedback voltage V_f, the output of the comparator V_A is a '0'. When V_x is smaller than V_f then V_A is a '1'. The timing generator produces mutually exclusive $n+2$ timing intervals of constant periods for an n-bit converter. It can be implemented by $n+2$

flip-flops interconnected as a shift register, and with an AND gate to set the first stage to '1', when all other stages are '0'.

The sequence control circuit is made of NAND gates. It produces SET and RESET signals for the storage circuits. The SET signals are generated by gating the $\overline{f'_c}$ with the timing intervals t_1 to t_n, whereas, the RESET signals are set by gating the $\overline{V_A \cdot f'_c}$ with the timing intervals t_1 to t_n.

For storage circuits, bistable latches (cross coupled NAND gates) can be used. Whenever any of the SET inputs is '0', the output is '1'. When any of the RESET inputs is '0', the output is '0'. The outputs of the storage circuits are inputted to the D-to-A converter and at the same time fed to the output gating logic.

There are n NAND gates which connect the outputs of the storage circuits to the output of the whole device during the timing interval t_{n+1}, i.e., during the clock period just after the one when the LSB has been tried. It is there to ensure that the output is obtained from the storage circuits only after the end of conversion, but before the storage circuits are RESET by the timing pulse t_{n+2}, since the outputs of the storage circuits are incomplete, or even in error, during all other times. The timing pulse t_{n+2} is fed simultaneously to the auxiliary reset inputs of the bistable latches in order to start the next conversion.

The switches in the $R/2R$ ladder D-to-A converter are driven by the outputs of the storage circuits. The D-to-A converter's output is fed back to the comparator, thus closing the loop.

Usually clock pulses are differentiated digitally in order to obtain narrow pulses which are then gated with V_A and timing pulses to produce SET and RESET pulses. Unlike the servo type A-to-D converter the successive approximation type A-to-D converter requires n clock periods to complete a conversion, regardless of the magnitude of the analogue input voltage, the conversion process is not continuous as for the servo type A-to-D converter. In this case the input voltage must remain constant during the conversion process, otherwise there will be an error in the binary output code at the end of the conversion. Because of this, it is a common practice to precede a successive approximation type converter with a sample and hold circuit, thereby holding the analogue input at a constant value during the conversion period.

The RS 427 is an 8-bit successive approximation A-to-D converter I.C. Three resistors and one capacitor are required to construct an accurate high speed A-to-D converter circuit suitable for many applications. The internal block diagram of the I.C. is shown in Fig. 7.14. Conversion is initiated by a Start Conversion (SC) pulse which sets the MSB to '1' and all other bits to '0'. The output data is valid when End of Conversion (EOC)

Fig. 7.14. Block diagrams of RS 427 type SA A-to-D converter. (RS Components Ltd.)

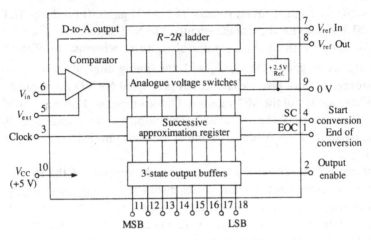

Fig. 7.15. Connection of external components for basic operation of RS 427 converter. (RS Components Ltd.)

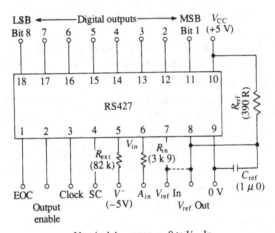

Nominal A_{in} range = 0 to V_{ref} In

goes HIGH and remains latched until the next SC pulse. Figure 7.15 shows the connection of external components for basic operation.

Worked example 7.7

A 4-bit successive approximation type A-to-D converter has a FSR of 10 volts. What are the bit values? If an analogue input signal of 8.3 volts is applied to the converter, what will be the binary output code? What will be the error? Draw a timing diagram for the conversion sequence.

Fig. ex.7.7

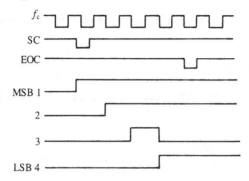

Solution

$$MSB = \frac{V_R}{2} = \frac{10}{2} = 5\,V$$

Next

$$MSB = \frac{V_R}{2^2} = 2.5\,V$$

Next

$$MSB = \frac{V_R}{2^3} = 1.25\,V$$

and

$$LSB = \frac{V_R}{2^4} = 0.625\,V.$$

The O/P code is 1101 which is equivalent to 8.125 V

$$\therefore \qquad error = \frac{8.3 - 8.125}{8.3} \times 100\%$$

$$= 2.11\%$$

Worked example 7.8

Calculate the time required by (i) a successive approximation type A-to-D converter and (ii) a servo type A-to-D converter to record a change in analogue voltage of 6.25 V. The frequencies of the clock pulses applied to the converters are 200 kHz and 500 kHz respectively. Both of them are 8-bit converters with FSRs of 10 V.

Solution

(i) The conversion time for the S.A. type of A-to-D converter is

$$\frac{1}{f_c} \times n = \frac{1}{200 \times 10^3} \times 8 = 40\,\mu s$$

(ii) The servo type A-to-D converter takes $\dfrac{1}{f_c}$ to record a change of 1 LSB equivalent, i.e.

$$\frac{1}{2^n} \cdot FSR = \frac{1}{2^8} \cdot 10 = \frac{10}{256}\,V$$

Therefore to record a change of 6.25 V it will take

$$6.25 \times \frac{256}{10} \times \frac{1}{f_c} = 6.25 \times \frac{256}{10} \times \frac{1}{500 \times 10^3} = 320\,\mu s$$

7.2.4   Dual slope A-to-D converter

This type of converter may be built at low cost for applications where speed of conversion is not an important factor. It is extensively used in precision digital voltmeters. Figure 7.16 shows the block diagram of a dual slope A-to-D converter. The switch S_2 resets the integrator prior to the start of each conversion. The input voltage may be kept constant with a sample and hold circuit during each conversion period. With the Start Conversion signal from the logic circuitry the switch S_2 opens and the switch S_1 connects the input analogue voltage to the integrator. The capacitor charges up linearly and at the end of a fixed period of time T_1, which can be set by the counter and control logic circuitry, the output voltage of the integrator becomes,

$$V_0(T_1) = -1/RC \int_0^{T_1} V_x dt = -V_x T_1/RC \qquad (7.12)$$

Fig. 7.16. Block diagram of dual slope A-to-D converter.

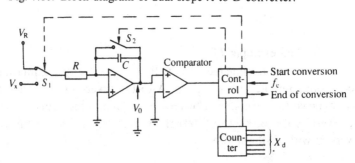

Fig. 7.17. Timing waveforms of dual slope A-to-D converter.

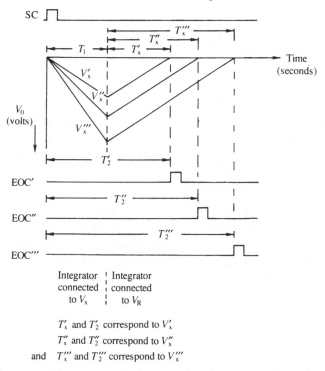

T'_x and T'_2 correspond to V'_x

T''_x and T''_2 correspond to V''_x

and T'''_x and T'''_2 correspond to V'''_x

Figure 7.17 shows the variation in the output voltage of the integrator with time for three different input voltages. By examining (7.12) we can see that the integrator output is directly proportional to the magnitude of the input voltage since T_1, R and C are constant. The integrator output is fed to a comparator whose reference voltage is zero. Thus as soon as the capacitor starts charging up the comparator produces a '1', and this after being gated with the clock pulses enables the counter to start counting. At the end of T_1 the logic circuitry makes the switch S_1 connect the integrator input to the reference voltage V_R which is of opposite polarity to that of the analogue signal input. Now the capacitor steadily discharges until the integrator output reaches zero. As soon as it does so, the comparator changes its output state and the logic circuitry stops the counter from counting. Suppose it takes T_2 seconds for the output voltage of the integrator to reach zero. From Fig. 7.17 we may write

$$V_0(T_2) = V_0(T_1) - 1/RC \int_{T_1}^{T_2} - V_R \, dt \qquad (7.13)$$

where $V_0(T_1)$ is the integrator output voltage at time T_1 and $V_0(T_2)$ is the integrator output voltage at time T_2.

Now substituting for $V_0(T_1)$ from (7.12) and making $V_0(T_2)=0$, we get from (7.13)

$$0= - V_x T_1/RC + V_R/RC(T_2 - T_1)$$

or

$$T_x = T_2 - T_1 = V_x/V_R T_1 \qquad (7.14)$$

Thus the time $T_x = (T_2 - T_1)$ is proportional to the analogue input voltage. Usually the count representing T_1 is chosen to be the full scale of the counter, therefore by reading the final output we can determine V_x. Again it may be possible to make the counter read V_x directly by choosing a proper value for V_R.

With moderately stable components the dual-slope A-to-D converter achieves very good accuracy which does not depend on the tolerances of the values of C and R since the RC product does not influence the counting time, as can be seen in (7.14). The input analogue signal is averaged during the measuring time, so any fluctuations in mains supply frequency or any high frequency noises are averaged out, which is of course another advantage.

Worked example 7.9

An 8-bit dual slope A-to-D converter has a clock frequency of 50 kHz. What is the conversion time for an input signal of 7.2 volts if the reference voltage is 10 volts? Assume that it takes 1.5 ms to zero the integrator.

Solution

conversion time $= T_1 + T_x +$ time to zero the integrator

$$= T_1 + V_x/V_R T_1 + \text{time to zero the integrator}$$

$$= (1 + 7.2/10)T_1 + 1.5 \, \text{ms}$$

but

T_1 is the time required to count full scale

or

$$T_1 = 2^n \times 1/f_c = 2^8 \times 1/50 \times 10^3$$

therefore the conversion time $= 1.72 \times 256/50 \times 10^3 + 1.5 \, \text{ms}$

$$= 8.81 + 1.5 \, \text{ms} = 10.31 \, \text{ms}$$

Fig. 7.18. Quantisation error.

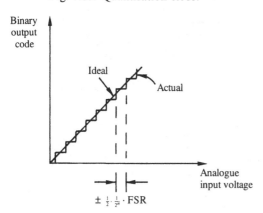

$\pm \frac{1}{2} \cdot \frac{1}{2^n} \cdot \text{FSR}$

7.3 Errors in D-to-A and A-to-D converters

As mentioned in previous sections, various factors such as the offset input current and voltage of the operational amplifiers acting as a comparator, the imperfect resistance ratios in D-to-A converters and other component imperfections give rise to errors and cause deviations from the ideal transfer characteristics of the converters shown in Figs. 7.2 and 7.10. In the following section different types of common errors are discussed.

(a) Quantisation error

If an analogue voltage is fed to an A-to-D converter and the output is subsequently inputted to a D-to-A converter, then ideally the latter will produce an analogue voltage exactly equal to the input analogue voltage. However, in practice the output analogue voltage may have a maximum deviation of $\pm \frac{1}{2}\text{FSR}/2^n$ from the input analogue voltage due to the fact that for an A-to-D converter each of the output binary codes represents a unique band of input voltage which is determined by the FSR and the number of bits of the device. By examining Fig. 7.18 it can be said that this type of error arises because the output of an A-to-D converter is only correct when the analogue input voltage is precisely equal to the analogue equivalent of the binary output code, and the maximum variation possible in input voltage without changing the binary code is $\pm \frac{1}{2}\text{FSR}/2^n$ which is equal to $\pm \frac{1}{2}$ the LSB weighting. This type of error known as the quantisation error is applicable to A-to-D converters but not to D-to-A converters. It is expressed as a percentage, e.g., ± 0.02 per cent FSR.

Fig. 7.19. Offset error.

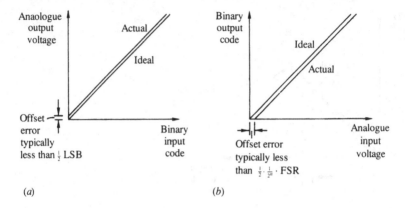

(a)

(b)

(b) Offset error

This, also called zero error, occurs in converters due mainly to the input offset current and voltage of the operational amplifiers used in the converters. In the case of D-to-A converters it is the value of the output analogue voltage when all the bits of the input binary code are zero. Conversely, in the case of A-to-D converters the offset error occurs when a zero input produces a binary output code other than the lowest, which is of course 00...0. Figures 7.19(a) and 7.19(b) show the transfer characteristics of D-to-A and A-to-D converters respectively with offset errors. Manufacturers often provide facilities to zero offset errors in high-resolution converters.

(c) Gain or scale-factor error

The gain or scale-factor determines the slope of the transfer characteristic of a converter. The gain or scale-factor error is, therefore, the difference between the slopes of the actual and ideal transfer characteristics of a converter when the offset error is zero. Figure 7.20 shows the transfer characteristics of A-to-D and D-to-A converters with gain errors.

Fig. 7.20. Gain or scale-factor error.

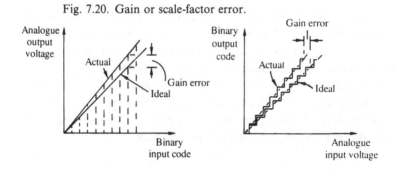

The reference voltage used in a converter partly determines its gain or scale-factor. In D-to-A converters and feedback type A-to-D converters the feedback resistor of the output operational amplifier contributes to the gain as well, whereas, in single-slope or dual-slope A-to-D converters the input resistor of the integrators does so. By adjusting all these variables the gain or scale-factor can be set to any desired value. The adjustment must be carried out after the offset error has been zeroed. This type of error is usually expressed as a percentage deviation from the ideal value.

(d) Linearity error

Ideally the transfer characteristics of converters are linear. For an ideal D-to-A converter a straight line may be drawn through all the D-to-A converter's output analogue voltages. For an ideal A-to-D converter the midpoints of each analogue voltage band may be joined together yielding a straight line. The linearity error occurs when the actual transfer characteristic of a converter deviates from the ideal characteristic as shown in Fig. 7.21. It is defined as the maximum deviation of the transfer characteristic from a straight line joining the zero and the full scale value, when the offset and gain errors have been reduced to zero. Unlike offset and gain errors which can be nulled to zero, linearity errors can not be adjusted, therefore its specification is quite important to the designer. This error is expressed in LSB or as a percentage of the FSR.

(e) Monotonicity error

This error occurs when the slope of the transfer characteristic of a D-to-A converter or feedback A-to-D converter changes sign as shown in Fig. 7.22 due to errors in the weighting of the bit current increments. It is the difference between the two adjacent output values of a converter at the step where the slope changes its sign.

Of the five types of errors mentioned above, the gain, offset and linearity errors are dependent on ambient temperature and power supply voltage.

Fig. 7.21. Linearity error.   Fig. 7.22. Monotonicity error.

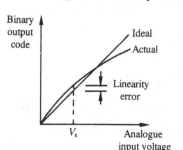

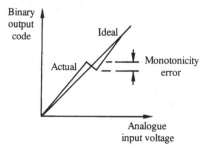

The values of these parameters also change as a function of time. How these various types of error contribute to the error of a total system, and hence affect its accuracy will be discussed later in this chapter.

Worked example 7.10
The RS DAC0800 used in the basic voltage output DAC system of Fig. 7.9 has the following error data: zero scale current (all bits off, I_{ZS}) $= 2\,\mu A$; non-linearity $= \pm 0.19\%$ FS; full-scale temperature coefficient $(TCI_{FS}) = \pm 50\,ppm/^{\circ}C$; power supply sensitivity (PSSI$_{FS+}$, PSSI$_{FS-}$) $= 0.01\%/\%$. The 777 type operational amplifier has the following data: $V_{io} = 5\,mV$; $\Delta V_{io}/\Delta T = 30\,\mu V/^{\circ}C$; $I_{os} = 40nA$ and $\Delta I_{os}/\Delta T = 10.3\,nA/^{\circ}C$. The reference voltage of 10 V has a tolerance of $\pm 1\%$ and a temperature coefficient (TCV$_{ref}$) of $\pm 25\,ppm/^{\circ}C$. The resistors have tolerances of 0.5% and both R_L and R_{REF} are 4.7 kΩ. Calculate the worst case errors before and after calibration. Assume that the maximum change in temperature is from 25 °C to 70 °C.

Solution
FSR

$$V_x = \frac{255}{256}\frac{V_{REF}}{R_{REF}} \times R_L \simeq \frac{10 \times 4.7 \times 10^3}{4.7 \times 10^3} = 10\,V$$

Worst case initial offset error

$$= \pm V_{io} \pm I_{os}R_L \pm I_{zs}R_L$$

$$= \pm 5 \times 10^{-3} \pm 40 \times 10^{-9} \times 4.7 \times 10^3 \pm 2 \times 10^{-6} \times 4.7 \times 10^3$$

$$= \pm 14.6\,mV$$

$$= \pm \frac{14.6 \times 10^{-3} \times 100}{10} = \pm 0.146\% \text{ of FS.}$$

The gain or scale factor is given by the ratio of the output voltage to the numerical fraction determined by digital input code.

$$\frac{V_x}{\left(\frac{1}{2} + \cdots \frac{1}{2^8}\right)} = I_{FS}R_L \cdot \frac{1}{\frac{255}{256}}$$

$$= \frac{255}{256} \cdot \frac{V_{REF}}{R_{REF}} \cdot R_L \cdot \frac{256}{255} = \frac{V_{REF}}{R_{REF}} \times R_L.$$

Thus the initial gain error comprises V_{REF}s tolerance, $\pm 1\%$, R_{REF}s and R_Ls tolerances of $\pm 0.5\%$ each, thus giving a worst case initial gain error $\pm 2\%$.

Linearity error is specified as $\pm 0.19\%$ of FS. Thus the worst case error without calibration

$$= \pm 0.146 \pm 2 \pm 0.19 = \pm 2.34\%.$$

After initial offset and gain errors are nulled to zero, the offset drift over the maximum temperature change of 45 °C ($70° - 25°$) [note: no data available for zero scale current-variation with temperature]

$$= \pm \frac{\Delta V_{io}}{\Delta T} \times 45 \pm \frac{\Delta I_{os}}{\Delta T} R_L \times 45$$

$$= \pm 30 \times 10^{-6} \times 45 \pm 10.3 \times 10^{-9} \times 4.7 \times 10^3 \times 45 = \pm 0.00352\ \text{V}$$

$$= \pm 0.035\%\ \text{FS}$$

The gain error over temperature change

$$\pm TCI_{FS} \times 100 \times 45 \pm TCV_{ref} \times 100 \times 45$$

$$= \pm \frac{50}{10^6} \times 100 \times 45 \pm \frac{25}{10^6} \times 100 \times 45 \simeq \pm 0.34\%$$

Linearity over temperature remains constant at 0.19%. Error over power supply variation $= \pm 0.01 \pm 0.01 = 0.02\%$. Thus the total worst case error after calibration

$$= \pm 0.035\% \pm 0.34\% \pm 0.19\% \pm 0.02\% = 0.59\%.$$

7.4   Multiplexers

An analogue multiplexer is a device which switches a number of different analogue input signals onto a single channel in a prescribed orderly fashion. The input channels can be accessed in a sequence or randomly by applying a proper digital address to the digital inputs. This device helps to time-share a single channel for many signals thus reducing the number of A-to-D converters, D-to-A converters and sample and hold circuits which otherwise may be needed for a particular data acquisition or distribution system. A multiplexer is basically an array of a number of analogue switches. At any particular instant only one switch is closed thus allowing the corresponding input signal to pass on to the single output channel. A 2^n channel multiplexer has 2^n input lines, one output channel and n control inputs. Figure 7.23 shows a simple 8 channel multiplexer. It needs a 3-bit ($2^3 = 8$) channel address. Channel addresses are inputed to the decoder from the counter in a cyclic order. The timing waveforms show the input and output waveforms for the multiplexer. The clock frequency must be chosen so that an input channel remains switched on to the output

Fig. 7.23. (a) A simple 8-channel multiplexer and (b) its input/output timing waveforms.

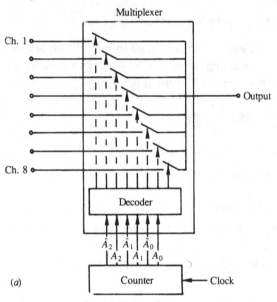

(a)

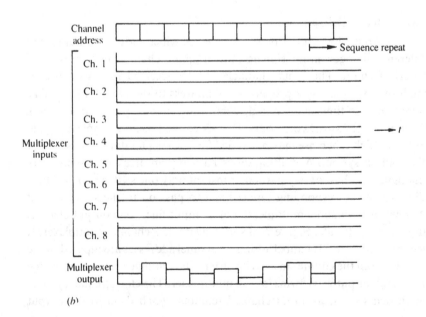

(b)

Fig. 7.24. Block diagram of a 16-channel multiplexer.

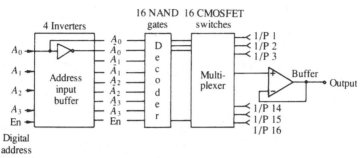

channel for a sufficient period in order to allow an A-to-D converter (which usually follows a multiplexer in a system) to complete the data conversion.

The block diagram of a 16 channel multiplexer is shown in Fig. 7.24. The 4-bit address (A_0-A_3) allows an access of 2^4, or 16 channels. The additional digital input known as the enable input, En, allows 'on–off' control of all the switches of the multiplexer. In systems where more than one multiplexer is required this enable input is essential. The address input buffer has four inverters and the decoder has sixteen 5-input NAND gates. These NAND gates are so connected that they produce a '0' (positive logic) at the outputs in a desired sequence. If $A_0 = A_1 = A_2 = A_3 = $ '0' and En = '1' then the first NAND gate will produce a '0'. All the other NAND gates will produce a '1' at their outputs. These will all simultaneously be applied to the respective switches.

Although there are several types of switches that are suitable for use in a multiplexer made of BJTs, JFETs or MOSFETs, the most commonly used one is the complementary MOSFET switch which is extremely small and inexpensive. The MOSFET is preferred to the JFET because the insulated gate of the MOSFET prevents the control voltage used to change the device's output state from entering the signal path. Since the 'on' resistance of a MOSFET changes as a function of the input voltage, switches with a complementary pair of MOSFETs as shown in Fig. 7.25 are used in the

Fig. 7.25. A complementary pair MOSFET switch.

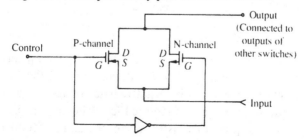

Fig. 7.26(a). V_{GS}–I_D and V_{DS}–I_D characteristics of a P channel MOSFET.

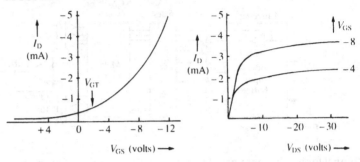

Fig. 7.26(b). V_{GS}–I_D and V_{DS}–I_D characteristics of an N channel MOSFET.

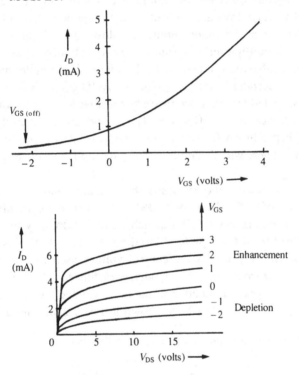

multiplexer. Figure 7.26(a) shows the V_{GS} against I_D and the V_{DS} against I_D characteristics of a P-channel MOSFET. It can be noted that the drain current of the device is enhanced by negative gate voltage and a negative gate-to-source threshold voltage is required for switching on the device. The V_{GS}–I_D and V_{DS}/I_D characteristics of an N-channel MOSFET are shown in Fig. 7.26(b). In this case a positive gate-to-source threshold

Fig. 7.27. Truth table for 16-channel multiplexer.

A_3	A_2	A_1	A_0	En	Channel 'on'
×	×	×	×	0	0
0	0	0	0	1	1
0	0	0	1	1	2
0	0	1	0	1	3
0	0	1	1	1	4
0	1	0	0	1	5
0	1	0	1	1	6
·	·	·	·	·	·
·	·	·	·	·	·
·	·	·	·	·	·
·	·	·	·	·	·
1	1	1	1	1	16

$\times = $ '1' or '0'

voltage is required to switch on the FET. Thus the channel resistance of each FET is a function of the polarity and amplitude of the input signal. However, because of the complementary action of the circuit, when the channel resistance of one FET decreases, the resistance of the other FET increases. Thus the effective resistance of the switch remains relatively constant for bipolar input signals of any amplitude. By examining the characteristics of the FETs we may say that large positive signals will pass through PMOS, whereas, large negative signals will pass through NMOS FETs. The 'on' resistance of the switch is typically $500\,\Omega$ and when the switch is off it presents a typical resistance of $50\,\text{M}\Omega$.

The '0' from the output of the first NAND gate of the decoder is applied to the gate of the P-channel MOSFET thus switching it on. At the same time the '0' output is inverted to a '1' and applied to the gate of the N channel MOSFET thus switching it on as well. Thus both the P- and N-channel MOSFETs are turned 'on' simultaneously producing very low resistances between the drains and the sources and in effect connecting the input of Channel 1 of the multiplexer to the single output channel. All other complementary MOSFET (CMOSFET) switches are off during this period. A truth table for a 16-channel multiplexer is shown in Fig. 7.27.

7.5 Demultiplexers

A demultiplexer takes an input and connects it to one of several possible outputs, according to an input digital address, while the other outputs are left open-circuited. The bilateral property of CMOSFET switches allows us to use them in demultiplexers as shown in Fig. 7.28. As in the case with multiplexers, in this case also the digital address can connect

Fig. 7.28. A basic 8-channel demultiplexer.

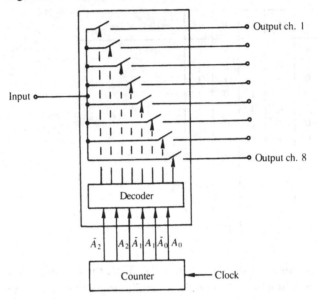

the input to the outputs in a cyclic order, or at random one at a time. When the input is not connected to an output channel, the output on that channel falls to zero. Therefore a sample-and-hold circuit is needed on each channel to hold up the voltage until replenished with the input voltage on the next cycle.

7.6. Errors in multiplexers

There are mainly two types of error which are generated in multiplexers. One is the static error and the other the dynamic error.

Static error

The static error in a multiplexer is caused principally by the 'on' resistance of the closed switch and the leakage currents which flow through all the other switches which remain open. Let us consider that at an instant the switch for the first channel is closed and the input voltage for that channel is V_{in}. The equivalent circuit for the multiplexer is shown in Fig. 7.29. It is apparent from the circuit diagram that the output voltage V_0 will not be the same as V_{in}, but differ by an amount $(N-1)I_{D(off)}r_{on}$ where N, $I_{D(off)}$ and r_{on} are the total number of channels of the multiplexer, the leakage current through the open switches and the 'on' resistance of the closed switch respectively. This difference between the input and output

Fig. 7.29. Equivalent circuit of a multiplexer in order to obtain the static error.

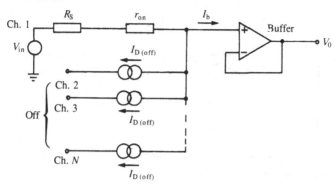

voltages is the static error of the multiplexer. If we take into account the bias current of the buffer which may be part of the next stage and assume that the bias current and the leakage currents are flowing in the same direction then the worst case static error can be obtained as follows,

$$(N-1)I_{D(off)}r_{on} + I_b r_{on} \tag{7.15}$$

where I_b is the bias current of the buffer.

So far we have not taken into account the voltage drop due to the channel on leakage current $I_{D(on)}$ from driver into the switch. Also if we consider the source resistance R_s then we may write the final expression for the static error as follows,

$$[(N-1)I_{D(off)} + I_{D(on)} + I_b](R_s + r_{on}) \tag{7.16}$$

where $I_{D(on)}$ is the 'channel ON leakage current' flowing from driver into the 'on' switch.

Dynamic error

The dynamic error occurs in a multiplexer due to the 'on' resistance of a closed switch and the various nodal and interelectrode capacitances which are present in the multiplexer. If we have an input which suddenly changes, then the output would be unable to follow the input instantaneously, but have an exponential rise with a time constant, τ determined by these capacitances and the 'on' resistance.

The magnitude of the dynamic error depends on the access time allowed for each input channel. The access time for a channel comprises an intentional delay time t_d which is introduced to avoid momentary shorting between two input channels and the exponential rise time t_e which is

Fig. 7.30. Access time for a channel of a multiplexer.

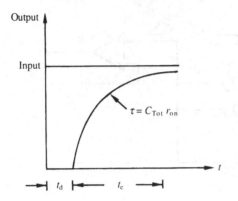

proportional to τ as illustrated in Fig. 7.30. The dynamic error is the difference between the magnitude of the input signal to be multiplexed and the maximum value transferred to the output of the device at the end of the access time.

Firstly let us consider the capacitances present in a multiplexer in order to understand fully the cause of the dynamic error. When a switch is open it can be represented by a network of capacitances as shown in Fig. 7.31. The source–body capacitance of each CMOSFET forms the channel input capacitance $C_{s(off)}$. It is the capacitance to ground looking into an 'off' switch. The drain–body capacitances of all the MOSFETs give rise to the multiplexer output capacitance $C_{D(off)}$ which is the capacitance seen looking into the output of the multiplexer. The drain–body capacitance of a single CMOSFET is typically 3 pF and since the output capacitance $C_{D(off)}$ is directly proportional to the number of analogue channels (off) of the multiplexer; for a 16-channel multiplexer it is typically 45 pF. The capacitance between the source and drain of a 'turned-off' switch is denoted by $C_{DS(off)}$. Its typical value is 1 pF and, therefore, smaller than the channel

Fig. 7.31. Capacitances across terminals and grounds for an open switch of a multiplexer.

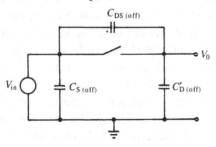

Fig. 7.32. Equivalent circuit of a multiplexer in which the open switches are replaced by capacitances.

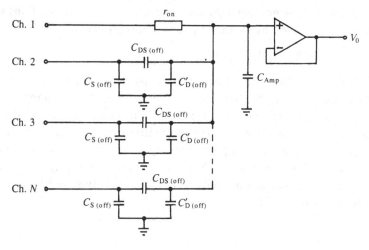

Fig. 7.33. Simplified equivalent circuit of a multiplexer.

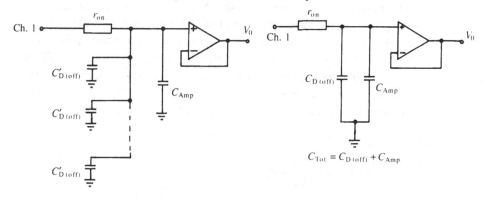

input capacitance and of course the multiplexer output capacitance. Figure 7.32 illustrates the equivalent circuit of a multiplexer in which the open switches have been replaced by the capacitive networks. The circuit can be simplified as displayed in Fig. 7.33 by neglecting the $C_{S(off)}$ and $C_{DS(off)}$.*

The time constant is, therefore, given by

$$\tau = r_{on} C_{Tot}$$

---

* The time constant is directly proportional to the effective capacitance between the 'on' resistance terminal and the ground. The capacitances $C_{S(off)}$ and $C_{DS(off)}$ form series circuits which are then in parallel with the $C'_{D(off)}$. Considering the typical values of the capacitances we can ignore the channel input capacitances and the source–drain capacitances.

where C_{Tot} is the sum of C_{Amp}, the input capacitance of the buffer and $C_{\text{D(off)}}$, the multiplexer output capacitance.

So far we have not taken the source resistance into account. If the source resistance is given by R_{S} then the time constant becomes

$$\tau = (R_{\text{S}} + r_{\text{on}})(C_{\text{Amp}} + C_{\text{D(off)}}) \tag{7.17}$$

The source–body and body–drain capacitances also give rise to the crosstalk between 'off' channels and an 'on' channel. The magnitude of the crosstalk is a function of the frequency and also the magnitude of the signal applied to the 'off' channel. It is given by the ratio of the magnitude of a test signal appearing at the output to the magnitude of the test signal applied at the input of an 'off' channel. The ratio is more commonly known as 'off isolation' and specified in decibels. The RS HI506 type multiplexer has an 'off isolation' of 75 dB.

Worked example 7.11

A multiplexer is part of a 12-bit, 10-volt system. The CMOSFET switches have leakage currents of 45 nA and 'on' resistances of 450 Ω. The multiplexer output capacitance is 44 pF. All inputs are scanned sequentially. (i) Calculate the number of input channels that the multiplexer has if the maximum static error is one-eighth of a least-significant-bit ($\frac{1}{8}$LSB). (ii) Find the maximum sampling rate per channel for a dynamic error equal to the static error. In addition to the settling time, a further 0.05 μsec must be allowed to avoid momentary shorting between input channels.

Solution

$$\frac{1}{8}\text{LSB} = \frac{1}{8} \times \frac{1}{2^n}\text{FSR}$$

$$= \frac{1}{8} \times \frac{1}{2^{12}} \times 10 = 0.000305 \text{ volts}$$

Say N is the number of channels, then the static error,

$$r_{\text{on}}(N-1)I_{\text{leakage}} = 0.000305$$

or

$$450(N-1) \times 45 \times 10^{-9} = 0.000305$$

or

$$N = 16$$

Now the settling time is given by

Fig. ex.7.11

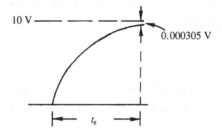

$$t_e = RC \ln \frac{10-0}{10-(10-0.000305)}$$

$$= 450 \times 44 \times 10^{-12} \times \ln 32786$$

$$= 205.8 \text{ nsec}$$

The access time is given by

$$t_A = t_d + t_e = (50 + 205.8) \times 10^{-9} = 255.8 \text{ nsec}$$

For 16 channels the access time is given by

$$t_A = 255.8 \times 10^{-9} \times 16 = 4092.8 \text{ nsec}$$

Therefore

$$\text{sampling rate/channel} = \frac{1}{4092 \cdot 8 \times 10^{-9}}$$

$$= 244300 \text{ samples per sec}$$

Worked example 7.12

In a data acquisition system 64 signals are scanned sequentially and are connected to an analogue-to-digital converter via nine multiplexers as shown in Fig. ex.7.12(a). Each one has eight input channels. The leakage

Fig. ex.7.12(a)

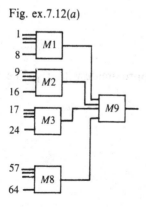

current is 10 nA when a switch is at 'off' state. The 'on' resistance of the switch r_{on} is 250 Ω. The output capacitance of each multiplexer is 50 pF. Calculate the static error and approximate settling time if a dynamic error of 0.1% is allowed.

Solution

At any particular moment only one channel of the multiplexer among the first eight is on in conjunction with one channel of the ninth multiplexer as shown in Fig. ex.7.12(*b*).

Fig. ex.7.12(*b*)

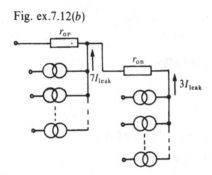

From the diagram,

$$\text{Static error} = r_{on} \times 7I_{leakage} + 2r_{on} \times 3I_{leakage}$$

$$= 13 r_{on} \times I_{leakage}$$

$$= 13 \times 250 \times 10 \times 10^{-9} = 0.032 \, \text{mV}$$

In order to find the settling time the capacitances have to be taken into account as shown in Fig. ex.7.12(*c*).

Fig. ex.7.12(*c*)

The time constant for this circuit is approximately $3r_{on}C_T$. Therefore

$$t_e \simeq 3r_{on}C_T \ln \frac{100-0}{100-(100-0.1)}$$

$$= 3 \times 250 \times 50 \times 10^{-12} \times 6.9$$

$$= 0.259 \, \mu\text{sec}$$

Worked example 7.13

Sixteen 16-channel multiplexers are used in a data acquisition system as shown in Fig. ex.7.13(a). The 'on' resistance of the switch is $270\,\Omega$ the capacitance across the open switch is 1 pF and the output capacitance of each multiplexer is 50 pF. To achieve an accuracy of 0.1% what would be the maximum channel frequency that can be processed with all channels having the same frequency bound? Also calculate the off isolation for the multiplexer at 250 kHz. The access time for each multiplexer is 300 nsec.

Fig. ex.7.13

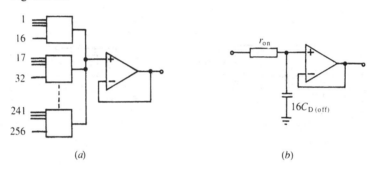

(a) (b)

Solution

The delay time is given by

$$t_d = t_A - t_e$$

$$= 300 \times 10^{-9} - 270 \times 50 \times 10^{-12} \times \ln\frac{100-0}{100-(100-0.1)}$$

$$= 207\ \text{nsec}$$

Therefore the total access time

$$t_A = t_d + t_e$$

$$= 207 \times 10^{-9} + r_{on} \times 16\,C_{D(off)} \ln\frac{100-0}{100-(100-0.1)}$$

$$= 207 \times 10^{-9} + 270 \times 16 \times 50 \times 10^{-12} \times 6.9$$

$$= 1697\ \text{ns}$$

Therefore the maximum sampling rate per channel is

$$\frac{1}{256 \times t_A} = \frac{1}{256 \times 1697 \times 10^{-9}} = 2302$$

According to the Nyquist sampling theorem a waveform is fully described by sampling its amplitude at a rate equal to twice the highest frequency present.

Therefore

$$2 \times f_{max} = 2302$$

or

$$f_{max} = 1151 \text{ Hz}$$

In order to find the off isolation, we consider the circuit shown.

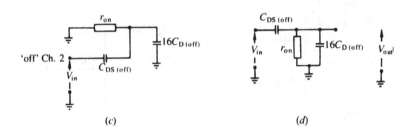

(c)   (d)

$$\frac{V_{out}}{V_{in}} = \frac{\dfrac{r_{on} \times \dfrac{1}{j\omega 16 C_{D(off)}}}{r_{on} + \dfrac{1}{j\omega 16 C_{D(off)}}}}{\dfrac{1}{j\omega C_{DS(off)}} + \dfrac{r_{on} \times \dfrac{1}{j\omega 16 C_{D(off)}}}{r_{on} + \dfrac{1}{j\omega 16 C_{D(off)}}}}$$

$$20 \log \left| \frac{V_{out}}{V_{in}} \right| = 20 \log \left| \frac{r_{on} j\omega C_{DS(off)}}{1 + j\omega r_{on}[C_{DS(off)} + 16 C_{D(off)}]} \right|$$
$$= 20 \log r_{on} \omega C_{DS(off)}$$

since

$$1 \gg \omega r_{on}[C_{DS(off)} + 16 C_{D(off)}]$$

$\therefore$ off isolation

$$= 20 \log 270 \times 2\pi \times 250 \times 10^3 \times 1 \times 10^{-12}$$

$$= -67 \text{ dB}$$

7.7   Sample-and-hold circuits

In earlier sections of this chapter the need of using a sample-and-hold circuit had been mentioned. Some A-to-D converters, such as the successive approximation type, should have their input voltages constant during the conversion periods otherwise erroneous results will occur. For a 12-bit 10-volt converter the variation in input during the conversion time should not be more than $\pm\frac{1}{2}\, 10/2^{12}$ for a conversion uncertainty of less than $\pm\frac{1}{2}\,\mathrm{LSB}$. If the conversion time for this particular A-to-D converter is $3\,\mu\mathrm{sec}$ then the maximum gradient of the input signal should be

$$\leqslant \frac{10}{2^{12}} \times \frac{1}{3 \times 10^{-6}}\,\mathrm{V\,s^{-1}} \tag{7.18}$$

If we consider a sinusoid of $\pm 5\,\mathrm{V}$ as an input signal, then its maximum slope is given by

$$\mathrm{d}V/\mathrm{d}t|_{\max} = 5.2\pi f$$

substituting for $\mathrm{d}V/\mathrm{d}t|_{\max}$ from (7.18) we get $f \leqslant 26\,\mathrm{Hz}$.

Thus a sample-and-hold circuit is needed in front of the A-to-D converter in order to maintain $\pm\frac{1}{2}\,\mathrm{LSB}$ accuracy for a signal frequency higher than $26\,\mathrm{Hz}$.

The output of a sample-and-hold circuit faithfully follows the signal present at the input upon command to 'sample', and upon a 'hold' command stores the input signal at the instant the 'hold' command is given. The sample and hold command signals are usually the logic '0' and logic '1'. The simplest form of sample-and-hold circuit is a switch and a capacitor as shown in Fig. 7.34(a). When the switch is closed the capacitor charges towards the applied input voltage at a rate determined by the time constant of the capacitor and the sum of switch resistance and the source resistance.

The simple circuit of Fig. 7.34(a) has a drawback that any load on the output will tend to discharge the capacitor. Therefore in practice a buffer is used after the capacitor as shown in Fig. 7.34(b). During the 'hold' period

Fig. 7.34(a). Simplest form of sample-and-hold circuit.
(b). Practical sample-and-hold circuit.

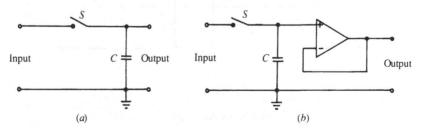

(a)   (b)

the output still may droop because of a finite bias current flowing into the operational amplifier. An N-channel MOSFET is used as the switch which can be operated electronically at a very high speed. The N-channel MOSFET needs a positive logic '1' to be turned on, thus connecting the capacitor to the input. The capacitor charging rate is determined by the time constant $C(R_s + r_{on})$ where R_s is the source resistance and r_{on} is the 'on' resistance of the MOSFET. When the MOSFET is on during the 'sample' period, the capacitor may cause a loading effect on the signal source by drawing a large current to charge itself. A buffer amplifier with sufficient output current may be used at the input to deliver a high charging current to the capacitor and also at the same time to provide a high impedance to the signal source. If the maximum current that the input buffer can supply is I_{max}, then the charging rate of the capacitor is given by

$$dV_c^+/dt = (I_{max} - I_b)/C \tag{7.19}$$

where I_b is the bias current for the operational amplifier. Generally $I_b \ll I_{max}$ and we may write

$$dV_c^+/dt = I_{max}/C \tag{7.20}$$

The circuit can be further modified by feeding the output of the device, in other words, the output of the second buffer back to the input buffer as illustrated in Fig. 7.35. It eliminates the offset error produced at the output of the previous circuit due to any offsets in the switch, input or output buffer amplifiers. In the modified circuit the feedback forces the capacitor to charge up at high speed to make the output of the sample-and-hold circuit equal to the input voltage. The input buffer amplifier is usually a high performance operational amplifier having an excellent slew rate. The

Fig. 7.35. Sample-and-hold circuit with MOSFET switch.

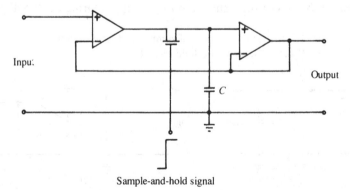

Sample-and-hold signal

output buffer amplifier has very high input impedance and needs extremely low bias current.

During the 'sample' period the MOSFET switch is on and the 'on' resistance of the MOSFET switch, r_{on} forms a low-pass filter in combination with the capacitor. So if high speed signals are to be followed accurately the capacitor has to have a small value. The maximum slew rate of the input voltage is determined by the maximum charging rate of the capacitor given by (7.20). Thus the slew rate of the whole sample-and-hold circuit depends on both I_{max} and r_{on}. During the 'hold' period when the sample-and-hold circuit is off, the bias current of the output buffer amplifier and the leakage currents of both the MOSFET switch and the capacitor cause the capacitor's voltage droop according to the expression

$$\mathrm{d}V_c^-/\mathrm{d}t = (I_b + I_{leakage(sw)} + I_{leakage(C)})/C \tag{7.21}$$

Therefore, the capacitor must be a precision capacitor, as large as possible, with a very low leakage current to minimise the droop.

Instead of using a single MOSFET as a switch in the circuit a complementary MOSFET switch can be used. This provides a low and constant 'on' resistance during the 'sample' period whatever the polarity and magnitude of the input signal is.

The MOSFET switches can also be replaced by a diode-bridge switch driven by a pulse transformer as shown in Fig. 7.36. The diode-bridge switches can be switched in a few hundred picoseconds. When the switch is enabled, all the four diodes are on and the 'on' resistance is that of a single diode usually of the order of tens of ohms. Typical sample-and-hold circuits using this type of switch have sampling rates of 1 MHz, sample-mode bandwidths of 45 MHz and output slew rates of 500 V μs^{-1}.

Fig. 7.36. Sample-and-hold circuit with diode-bridge switch.

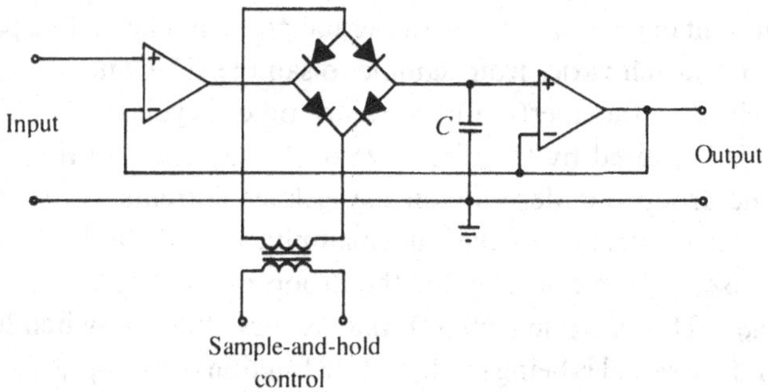

Sample-and-hold
control

Fig. 7.37. Ideal sample-and-hold operation.

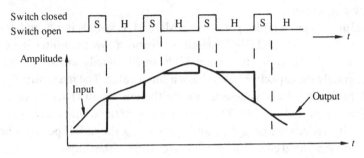

Performance parameters

Figure 7.37 illustrates the perfect sample-and-hold operation, but in reality there are some limitations in the performance of the sample-and-hold circuits as we will discuss now.

Input impedance. This is the impedance presented to the analogue input signal and depends greatly on the type of circuit used. In circuits with input buffer amplifiers the input impedances are of the order of tens of MΩ, but in very high speed sample-and-hold circuits where input bufferings are omitted, the input impedance has one value when the switch is closed and another value when it is open.

Acquisition time. This is the time required, after the sampling switch is closed, for the capacitor to charge to a full-scale voltage change and then remain within a specified error band around the final value. It is mainly a function of capacitor charging currents, time constants of the charging circuit and the slew rates of the buffer amplifiers. Typical values are 50 ns–10 μs.

Aperture time. This consists of two parts: (i) the aperture delay time and (ii) the aperture uncertainty time. The former is the period between the receipt of the hold command and the opening of the sampling switch. It is typically 10 nsec. Due to sampling switch characteristics, the aperture time contains a small amount of uncertainty, i.e. the actual point in time of the opening of the sampling switch varies from sample to sample. This time variation, or jitter, specified as the aperture uncertainty time is typically 30–200 psec.

Droop. This is caused by the discharge of the capacitor during the hold period. The droop rate depends on the leakage currents in the sampling switch, the bias currents of the buffer amplifiers and the leakage of the capacitor itself. A typical value for the droop rate is 50 μV s^{-1}.

Feedthrough. This is a phenomenon that occurs after the switch has been opened and the signal is being held. A small fraction of the input signal feeds

Fig. 7.38. Practical sample-and-hold operation and performance parameters.

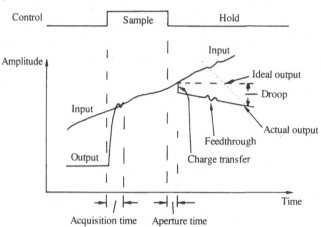

through the open-switch interelement capacitance. Since the coupling is capacitive, it is a function of frequency and expressed in terms of dB over a band of frequencies. A typical value is $-50\,\text{dB}$ at $10\,\text{MHz}$.

Charge transfer. When the switch is opened a small charge due to the capacitance of the switch terminal is transferred to the output terminal together with the charge of the capacitor. It causes the output to drop as illustrated in Fig. 7.38.

Gain. This is the ratio of the output voltage to the input voltage at the instant the switch opens. It is usually less than unity because of the feedback in the circuits. A typical value is 0.95.

Linearity. This is the maximum amount by which the input–output characteristic of the sample-and-hold circuit deviates from a straight line during the sample period. The linearity error is normally specified as a percentage of the full-scale, a typical value being $\pm 0.25\%$ maximum for input signal changes of $\pm 1.25\,\text{V}$.

Offset. This is defined as the amount of output voltage when the sampled input voltage is zero. It is expressed in millivolts or as a percentage of full-scale. However, since the d.c. offset can be adjusted to zero, manufacturers do not specify it in data sheets, but the variation in offset with temperature is an important factor and a typical value for output offset voltage drift is $50\,\mu\text{V}/^\circ\text{C}$.

Figure 7.38 shows the effects of dynamic parameters in exaggerated form. Effects of all these are expected to be less than $\frac{1}{2}$ LSB of the A-to-D converter to which it may be connected, otherwise the main purpose of using a sample-and-hold circuit would not be served.

Worked example 7.14

The circuit in Fig. ex.7.14 is connected to a source with an output impedance of 600 Ω. The 'on' resistance of the FET switch is 250 Ω. The bias current of the operational amplifier is 50 pA. Find the acquisition time to 1% for a 10 V change in the output if the value of the capacitor is 0.01 μF. Find also the droop rate of the output voltage when the circuit is in hold mode.

Fig. ex.7.14

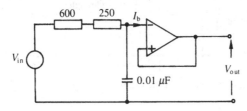

Solution

$$R = R_s + r_{on}$$

$$t_{acq} = RC \ln \frac{10 - 0}{10 - (10 - 0.1)} = (600 + 250) \times 0.01 \times 10^{-6} \ln 100$$

$$= 39.1 \, \mu sec$$

When the switch is open

$$\text{droop rate } \frac{dV_{out}}{dt} = \frac{I_b}{C} = \frac{50 \times 10^{-12}}{0.01 \times 10^{-6}} = 5 \, mV/s$$

Worked example 7.15

The sample-and-hold circuit in Fig. ex.7.15 has the following data: $C = 0.1 \, \mu F$, maximum output current of the operational amplifiers $= 100 \, mA$, amplifier bias currents $= 10 \, nA$. It is used in a 10 V, 10-

Fig. ex.7.15

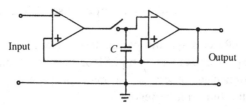

bit system. Calculate the acquisition time and the droop rate. If the sample-and-hold circuit is connected to an A-to-D converter what must be the ADC's maximum permitted conversion time; the largest permissible error being less than 1/8 LSB?

Solution
From (7.19)

$$\frac{dV_c^+}{dt} = \frac{I_{max} - I_b}{C}$$

$$\int_0^{10} dV_c = \frac{I_{max} - I_b}{C} \int_{t_1}^{t_2} dt$$

neglecting I_b since $I_b \ll I_{max}$

$$\therefore \quad t_2 - t_1 = \frac{C}{I_{max}} \times 10 = \frac{0.1 \times 10^{-6} \times 10}{100 \times 10^{-3}} = 10\,\mu s$$

From (7.21) ignoring $I_{leak(sw)} + I_{leak(c)}$ droop rate

$$= \frac{dV_c^-}{dt} = \frac{I_b}{C} = \frac{10 \times 10^{-9}}{0.1 \times 10^{-6}} = 100\,mV/s$$

$$\frac{1}{8}LSB = \frac{1}{8} \cdot \frac{1}{2^n} \cdot FSR = \frac{1}{8} \cdot \frac{1}{2^{10}} \cdot 10 = 0.00122\,V$$

100 mV drops in 1 s.

$\therefore$ 1.22 mV drops in 0.0122 s

so, the maximum conversion time is 12.2 ms.

7.8   System error analysis

       The total accuracy of a data acquisition or distribution system is a function of all the different error sources within it. These error sources confine themselves to the elements in the system. If the inaccuracies or error contributions of each stage in the system are considered separately, their peculiar error sources would not seem particularly difficult, however, studying the combined effect of all these individual sources can be very complex. Hence a practical approach for error analysis must be adopted.

       Two methods which are usually employed to analyse the errors in data acquisition systems are the worst-case method and the total-statistical approach. The first method assumes that all errors occur at their maximum, worst-case value, in the same direction and at the same time; so the total error is the algebraic sum of the worst-case values. However, the above assumption is highly improbable, since the error sources are affected by a number of independent factors. The second method assumes that all system parameters are statistically independent and behave randomly with respect to one another. It further assumes that each parameter is a normally distributed random variable. The total error due to all parameters calculated using the second method is given by

$$\text{Total error} = \left[\sum_{n=1}^{n} \varepsilon_n^2 \right]^{1/2} \tag{7.22}$$

where ε_n is the individual error source.

In order to calculate the system error it is appropriate to categorise the different types of error encountered in the system into three classes:

(i)   average errors: these are the nominal values of parameters such as offset, bias and leakage. These are added algebraically.

(ii) systematic errors: these always vary as some definite function of the system operating conditions such as temperature and power supply. All systematic errors due to specific causes are added algebraically to form parametric error functions.

(iii) random errors: these are mainly due to a random selection of a manufacturer's components. All the random errors are added according to (7.22).

The total system error can be obtained by combining the above three errors in the following manner:

$$\text{Total error} = \sum_{k=1}^{k} \varepsilon_{ak} + \left[\sum_{l=1}^{l} \varepsilon_{pl}^2 + \sum_{m=1}^{m} \varepsilon_{rm}^2 \right]^{1/2} \tag{7.23}$$

where

> ε_{ak} is the average error due to cause k,
>
> ε_{pl} is the parametric error due to cause l

and

> ε_{rm} is the random error due to cause m.

Although the quantisation error due to a converter is considered as a random error, it is usually shown as a separate factor together with (7.23).

Worked example 7.16

In order to analyse a complex waveform the data acquisition system shown in Fig. ex.7.16 is used. The error sources and average errors in percentage are given in Table 7.1. The systematic errors due to

Table 7.1

Elements	Average errors %	Random errors %	Systematic error	
			Temperature $\Delta = \pm 25\,^{\circ}\text{C}$	Power supply $\Delta = \pm 5\%$
1 Amplifier A_1:				
Bias current	+0.005	±0.01	±0.01	±0.01
Offset voltage	+0.05	±0.01		
Noise	+0.048			
Amplitude error	+1.9	±0.04	±0.005	
Gain non-linearity	+0.437			
2 RC LPF:				
Amplitude error	0.5	±0.05	±0.01	
3 Multiplexer:				
Leakage current	−0.0015	±0.02	±0.02	±0.005
Offset voltage	0.0014	±0.01		
4 Amplifier A_2:				
Bias current	0.001	±0.01		±0.01
Offset voltage	0.001	±0.01	±0.01	
Noise	0.001			
Amplitude error	1.3	±0.01	±0.01	
Gain non-linearity	0.0052			
5 Band-pass filter:				
Amplitude error	1.7	±0.01	±0.01	
Crosstalk	0.006			
Feedthrough	0.002			
6 A-to-D converter:				
Bias current	0.007	±0.01	±0.01	±0.01
Offset voltage	−0.03	±0.01		
Non-linearity	0.39			
Quantization error	0.2			

Fig. ex.7.16

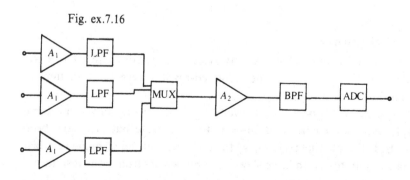

temperature change and power supply variation can be neglected. Assuming that the random errors are approximately 10% of the average errors, compute the system accuracy and express it in the form of (7.23).

Solution

Add the average errors algebraically.

$$0.005 + 0.05 + 0.048 + 1.9 + 0.437 + 0.5 - 0.0015 + 0.0014 + 0.001$$
$$+ 0.001 + 0.001 + 1.3 + 0.0052 + 1.7 + 0.006 + 0.002 + 0.007$$
$$- 0.03 + 0.39 = 6.32\%$$

Add all systematic errors due to specific causes algebraically.

Due to temperature: –

$$\varepsilon_{PT} = \pm 0.01 \pm 0.005 \pm 0.01 \pm 0.02 \pm 0.01 \pm 0.01 \pm 0.01 \pm 0.01$$
$$= \pm 0.085\%$$

Due to power supply variation:

$$\varepsilon_{pps} = \pm 0.01 \pm 0.005 \pm 0.01 \pm 0.01 = \pm 0.035\%$$

Root sum square all random errors with the parametric systematic errors: –

$$\sqrt{[(0.085)^2 + (0.035)^2 + (0.01)^2 + (0.01)^2 + (0.04)^2 + (0.05)^2}$$
$$+ (0.02)^2 + (0.01)^2 + (0.01)^2 + (0.01)^2 + (0.01)^2 + (0.01)^2$$
$$+ (0.01)^2 + (0.01)^2]$$
$$= \sqrt{[0.01385]} = 0.117\%$$

Total error $= 6.32\% \pm 0.117\% \pm 0.2\%$

↑   ↑   ↑

average random quantisation
and
systematic

Summary

1. Two types of D-to-A converters have been discussed. The R–$2R$ ladder type is preferred to the weighted-resistor type for thin film or monolithic IC technology.

2. Principles of operation of several types of A-to-D converters, some simple and inexpensive, and some not so accurate but fast have been described. The parallel type is easy to construct and can be made very fast using comparators with large slew rates, but suffers from poor resolution, whereas the successive approximation type is of complex nature and takes a

finite time for conversion whatever the magnitude of the change of input signal, but has a very good accuracy. For continuously but slowly varying signals the servo type is the best in respect of cost, resolution and conversion time. Another type which has been discussed in this chapter is the dual slope A-to-D converter. It is rather slow, but converts the average of an input signal thus removing any low frequency noises from the input.

3. The digital code consists of several bits which have weightings in a binary sequence. The bit having the smallest weighting is called the least significant bit. It is related to the analogue voltage reading as follows:

$$LSB = \frac{1}{2^n} \cdot FSR$$

where n is the number of bits in the digital code.

4. Among various types of error which occur in converters, the gain, offset and linearity errors are dependent on ambient temperature and power supply voltage. The first two types of error can be nulled to zero, whereas the linearity error cannot be adjusted. The errors are expressed in terms of a least significant bit (LSB) or percentage of the full-scale reading (FSR). Since the errors are mostly less than $\pm \frac{1}{2}$LSB, the converters with larger numbers of bits produce smaller errors.

5. Multiplexers are arrays of electronic switches (made of complementary MOSFETs) which connect a number of different analogue input signals onto a single channel in a sequence or at random thus reducing the requirements of other devices in a data acquisition or distribution system. Due to the bilateral property of complementary MOSFET switches the multiplexers can also be used as demultiplexers which connect one input to several output channels in a sequence or at random.

6. The sampling rate of the multiplexers and demultiplexers depends on the time required to access the input signals. The access time in turn depends on the 'on' resistance of the switches and the capacitances between the terminals of the 'off' switches and the ground.

7. In order to keep the input of some converters, such as the successive type A-to-D converter, constant during the conversion period to avoid any error, sample-and-hold circuits are absolutely necessary. Sample-and-hold circuits are also used as deglitchers. Among many parameters the most important ones are the settling time and droop rate. These should be as small as possible and determine the component values of a circuit.

8. The total error in a system comprising various circuits can be determined using two methods. One is the worst case method and the other the total statistical approach. The accuracy for a system incorporating most of the devices described in this chapter has been determined in an example.

Problems

7.1. A six-bit weighted-resistor D-to-A converter uses a Zener diode having a breakdown voltage of 2.5 volts. If the smallest weighted-resistor has a value of 15 kΩ and the feedback resistance is 30 kΩ, what is the FSR of the converter? What would be the output of the converter if the binary input is 100110?

7.2. The smallest resistor in the weighted-resistor D-to-A converter is 20 kΩ with a tolerance of ±1%. The 'on' resistance of the FETs acting as switches is 350 Ω. Calculate the accuracy of the converter. Neglect the other factors which may also affect the accuracy.

7.3. The output of the ladder type digital-to-analogue converter shown in Fig. 7.6 is connected to an analogue meter having a FSR of 15 volts. What is the value of R_2 if the resistors used in the ladder network are of 10 kΩ and 20 kΩ values? The reference voltage for the circuit is −10 volts.

7.4. A digital input 10011001 is applied to an 8-bit RSDAC0800 type D-to-A converter. Find the output current if the reference current of the converter is 5 mA. What is the value of the complementary output current $\bar{I}_0$?

7.5. A sinusoidal signal, $1.5 \sin 2\pi \times 15 \times 10^3 t$ is applied to the input of an 8-bit servo type analogue-to-digital converter having a FSR of 12 volts. What is the minimum frequency of the clock pulses?

7.6. For a 12-bit servo type A-to-D converter, the full-scale voltage is 10 volts and the clock frequency is 1 MHz. What is the largest permitted rate of change of the input voltage?

7.7. The FSR of a 6-bit successive approximation type A-to-D converter is 9 volts. What are the bit values? If an analogue input of 6.9 volts is applied to the converter, what will be the error at the output? Draw a timing diagram for the conversion sequence.

7.8. Which of the two, (i) a successive approximation type or (ii) a servo type A-to-D converter, will be faster to record a change in analogue voltage of 3.79 volts? The frequency of the clock pulses applied to the converters are 100 kHz and 250 kHz respectively. Both of them are 6-bit converters with FSR of 9 volts.

7.9. A 10-bit dual slope A-to-D converter has a reference voltage of 9 volts and a clock frequency of 100 kHz. What is the conversion time for an input signal of 6 volts assuming that it takes 1 ms to zero the integrator?

7.10. A D-to-A converter used in the system shown in Fig. 7.9 has the following error data: zero scale current = 1 μA; non-linearity = $\pm$0.1% FS; full scale temperature coefficient = 30 ppm/°C; power supply sensitivity = 0.01%/%. The associated operational amplifier has the following data: V_{io} = 3 mV; $\Delta V_{io}/\Delta T$ = 25 μV/°C; I_{os} = 45 nA and $\Delta I_{os}/\Delta T$ = 12 nA/°C. The reference voltage of 10 V has a tolerance of $\pm$0.5% and both R_L and R_{REF} are 6.8 kΩ. Calculate the worst case errors before and after calibration. Assume that the maximum change in temperature is from 15 °C to 60 °C.

7.11. The CMOSFET switches of a 12-bit, 9 V full-scale multiplexer have leakage currents of 30 nA and 'on' resistances of 250 Ω. If the number of input channels of the multiplexer is eight, calculate the static error.

7.12. The output capacitance of the multiplexer of Prob. 7.11 is 35 pF. Find the maximum sampling rate per channel for a dynamic error of 0.25 mV. All inputs are scanned sequentially. In addition to the settling time, a further 0.03 μs must be allowed to avoid momentary shorting between input channels.

7.13. In a data acquisition system thirty-two signals are scanned sequentially and are connected to an analogue-to-digital converter via five multiplexers as shown in Fig. P.7.13. The leakage current is 8 nA when a switch is at 'off' state. The 'on' resistance of the switch is 350 Ω. The output capacitance of each multiplexer is 35 pF. Calculate the static error and the settling time if a dynamic error of 0.25% is allowed.

Fig. P.7.13

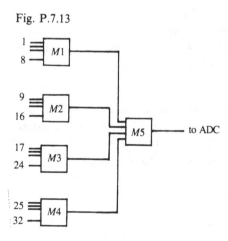

7.14. Twelve 8-channel multiplexers are used in a data acquisition system as shown in Fig. P.7.14. The 'on' resistance of each switch is 300 Ω, the capacitance across the open switch is 1 pF and the output capacitance of

Fig. P.7.14

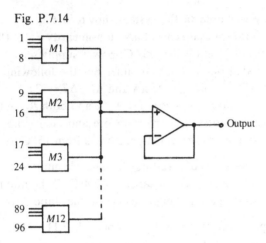

each multiplexer is 35 pF. In order to achieve an accuracy of 0.2%, what would be the maximum channel frequency that can be processed with all channels having the same frequency bound? The access time for each multiplexer is 250 ns.

7.15. A 10-bit successive approximation type A-to-D converter with conversion time of 20 μs is used on a signal, $v_{in} = 2(1 + \cos 2\omega t)$ volts, ($\omega = 2\pi f$, $f = 100$ Hz). The full-scale input voltage is 10 volts. Is a sample-and-hold circuit needed before the ADC?

7.16 A sample-and-hold circuit is connected to a source with an output impedance of 600 Ω as shown in Fig. P.7.16. The 'on' resistance of the FET switch is 370 Ω. The bias current of the operational amplifier is 80 pA. Find the acquisition time to 0.5% for a 9 V change in the output if the value of the capacitor is 22 nF. What would be the droop rate of the output voltage in 'hold' mode?

7.17. The sample-and-hold circuit shown in Fig. ex.7.15 is used in a 9 V, 12-bit system and has the following data: $C = 47$ nF, maximum output current of the operational amplifiers $= 150$ mA and the amplifier bias currents $= 40$ nA. Calculate the acquisition time and the droop rate. If the

Fig. P.7.16

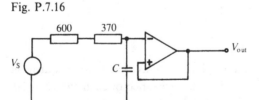

circuit is connected to an A-to-D converter, what must be its maximum conversion time, the largest permissible error being less than $\frac{1}{2}$ LSB?

7.18 The sample-and-hold circuit in Fig. P.7.18 is connected to the input of a 12-bit, 9 V FSR successive-approximation type A-to-D converter which has got a conversion time of 60 μs. The circuit has the following data: maximum output current of the operational amplifiers = 100 mA, amplifier bias currents = 10 nA, leakage current of the switch = 40 nA. The acquisition time is to be $\leqslant 1$ μs. All other effects are negligible. Calculate the maximum and minimum values for C which ensure satisfactory operation.

Fig. P.7.18

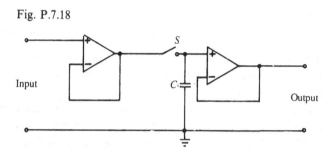

8

Computer aided circuit design

Objectives

At the end of the study of this chapter a student should be:

1. familiar with the modelling of different types of active devices
2. able to develop a suitable model of a device for a particular computer analysis
3. familiar with nonlinear d.c. analysis, small-signal a.c. analysis and large-signal transient analysis of simple electronic circuits
4. familiar with the Newton–Raphson algorithms and be able to determine the d.c. operating points of circuits for further analysis, and
5. familiar with the applications of some computer programs.

In recent years computational methods have been very popular for analysing and designing electronic circuits. It is now possible to design integrated circuits having thousands of transistors on a single chip. Such designs cannot be carried out experimentally at the bench. As very large scale integrated circuits make the fabrication of faster and cheaper computers possible, computer aided design is being used more and more to build such circuits. In this chapter, we will discuss various transistor models and parameters needed for computer analysis, different types of analysis and computer programs.

8.1. Computer aided design models

Renewed interest in transistor modelling took place in the sixties and seventies with the advent of computer aided design. Various models which had been developed during this period fell into two main categories, the first being single lump models in which transistor terminal currents are described in terms of quantities determined from terminal measurements; the second category of models is completely described by five basic transistor equations (two continuity equations, two current density equations and Poisson's equation) derived from the donor and acceptor concentration pattern. One has to be careful in choosing the right model for computer aided design, because an over-simplified model may produce inaccurate results, whereas, use of a model more complicated than necessary will obviously increase computer time. In this chapter we will consider mainly the most popular Ebers–Moll model which has been modified over the years to include various effects. One of the advantages of using this particular model is that the parameters describing the model can be easily measured.

8.1.1   Ebers–Moll models of bipolar transistors

The physical configuration of a junction transistor is that of two p–n junction diodes back-to-back, with a thin n-type or p-type region between them. Figure 8.1 shows the physical structure of an n–p–n transistor made by diffusing p-type and n-type impurities into the n-type material. The thin p-type region separating the two n-type regions, is referred to as the base, whereas the n-type regions are referred to as the emitter and collector.

Each of the emitter and collector currents is made up of two independent components. One component of the emitter current corresponds to minority-carrier injection or extraction at the emitter junction and is controlled by the base-emitter voltage. The other component of the emitter current is due to the transport of the minority-carriers across the base and is controlled by the base-collector voltage. These two components of the emitter current can be expressed as follows:

$$I_{EF} = I_{ES}(e^{qV_{BE}/kT} - 1) \tag{8.1}$$

$$I_{ER} = \alpha_R I_{CR} \tag{8.2}$$

where

I_{ES} is the emitter-base saturation current,

V_{BE} is the base-emitter voltage,

α_R is the common-base reverse current gain

Fig. 8.1. Physical structure of an n–p–n transistor.

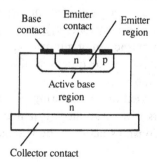

and

q, k and T are electron charge, Boltzmann's constant and absolute temperature respectively. α_R may vary from near unity to very much less than unity.

Similarly one component of the collector current corresponds to minority-carrier injection or extraction at the collector junction and is controlled by the base-collector voltage, whereas, the other component results from transport of the minority-carriers across the base and is controlled by the base-emitter voltage. These two components of the collector current are given by:

$$I_{CR} = I_{CS}(e^{qV_{BC}/kT} - 1) \tag{8.3}$$

$$I_{CF} = \alpha_F I_{EF} \tag{8.4}$$

where

I_{CS} is the collector-base saturation current,
V_{BC} is the base-collector voltage

and

α_F is the common-base forward current gain. α_F usually lies in the range 0.9 to 0.995.

The subscripts F and R refer to the forward and reverse components of currents controlled by the base-emitter and base-collector voltages respectively. We may combine (8.1)–(8.4) in order to obtain the total emitter and collector currents of an n–p–n transistor as follows,

$$I_E = -I_{EF} + I_{ER} = -I_{ES}(e^{qV_{BE}/kT} - 1) + \alpha_R I_{CS}(e^{qV_{BC}/kT} - 1) \tag{8.5}$$

$$I_C = +I_{CF} - I_{CR} = \alpha_F I_{ES}(e^{qV_{BE}/kT} - 1) - I_{CS}(e^{qV_{BC}/kT} - 1) \tag{8.6}$$

These equations have been derived by assuming an ideal transistor of uniform cross-sectional area and neglecting the recombination of carriers in the base region. The Ebers–Moll model shown in Fig. 8.2 represents these expressions with the aid of idealised junction diodes and dependent current sources. The diodes have saturation currents I_{ES} and I_{CS}, and account for the processes of injection or extraction of minority-carriers at either junction. The current sources account for the processes of transport of minority-carriers across the base region.

In this chapter analyses have been carried out considering only n–p–n transistors, however, similar treatments can be applied to p–n–p transistors. For a p–n–p transistor the relationships between the voltages and the currents are given by

Fig. 8.2. Ebers–Moll model of an n–p–n transistor controlled by diode currents or terminal voltages.

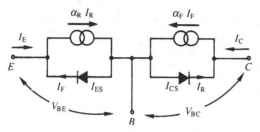

$$I_E = I_{ES}(e^{qV_{EB}/kT} - 1) - \alpha_R I_{CS}(e^{qV_{CB}/kT} - 1) \tag{8.7}$$

$$I_C = -\alpha_F I_{ES}(e^{qV_{EB}/kT} - 1) + I_{CS}(e^{qV_{CB}/kT} - 1) \tag{8.8}$$

Sometimes it is desirable to transform the Ebers–Moll model to a form where the dependent sources are controlled by 'terminal currents' instead of 'diode currents'. The required expressions for an n–p–n transistor can be obtained from (8.5) and (8.6) and are given by,

$$I_E = -\alpha_R I_C - I_{EO}(e^{qV_{BE}/kT} - 1) \tag{8.9}$$

$$I_C = -\alpha_F I_E - I_{CO}(e^{qV_{BC}/kT} - 1) \tag{8.10}$$

where I_{EO} is the reverse-biased emitter current measured with the collector terminal open and can be expressed as

$$I_{EO} = I_{ES}(1 - \alpha_F \alpha_R). \tag{8.11}$$

Similarly, I_{CO} is the reverse-biased collector current with the emitter terminal open and can be expressed as

$$I_{CO} = I_{CS}(1 - \alpha_F \alpha_R). \tag{8.12}$$

Figure 8.3 shows a model which is controlled by terminal currents. It is completely interchangeable with the one controlled by diode currents or terminal voltages shown in Fig. 8.2.

Fig. 8.3. Ebers–Moll model of an n–p–n transistor controlled by terminal currents.

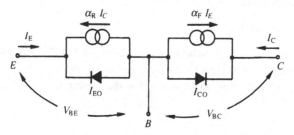

Fig. 8.4. Transport version of the Ebers–Moll model for an n–p–n transistor.

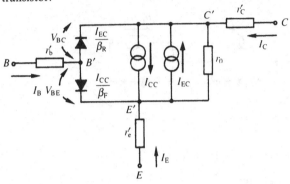

Development of the model

DC model. There are two versions of the Ebers–Moll model: the Injection version and the Transport version. In the former case, which we have just discussed, the reference currents used in the representation of the dependent parameters are the diode currents injected at the junctions, whereas, in the latter case the reference currents are those traversing the base region. Since the functional dependence of the model parameters in the latter version are more realistic from the physical point of view and at the same time simplify the measurement techniques of the parameters, we will use this version to develop the model.

The transport version of the Ebers–Moll model for an n–p–n transistor is shown in Fig. 8.4. The reference current at the collector for this model is given by

$$I_{CC} = I_S[\exp(qV_{BE}/kT) - 1] \tag{8.13}$$

and the reference current at the emitter is given by

$$I_{EC} = I_S[\exp(qV_{BC}/kT) - 1] \tag{8.14}$$

and

where the transistor saturation current,

$$I_S = \alpha_F I_{ES} = \alpha_R I_{CS} \tag{8.15}$$

β_F and β_R are the common-emitter large-signal forward and reverse current gains respectively and are given by $\beta_F = \alpha_F/(1 - \alpha_F)$ and $\beta_R = \alpha_R/(1 - \alpha_R)$. From Fig. 8.4 we may find expressions for the terminal currents for the transport model,

$$I_E = \frac{-I_{CC}}{\beta_F} - I_{CC} + I_{EC} = -I_{CC}\left(1 + \frac{1}{\beta_F}\right) + I_{EC} \tag{8.16}$$

Now substituting for I_{CC}, I_{EC} and I_S into (8.16):

$$I_E = \alpha_F I_{ES}\left[-(e^{qV_{BE}/kT} - 1)\left(1 + \frac{1-\alpha_F}{\alpha_F}\right) + (e^{qV_{BC}/kT} - 1) \right]$$

$$= \alpha_F I_{ES}\left[-\frac{1}{\alpha_F}(e^{qV_{BE}/kT} - 1) + (e^{qV_{BC}/kT} - 1) \right]$$

$$= -I_{ES}(e^{qV_{BE}/kT} - 1) + \alpha_R I_{CS}(e^{qV_{BC}/kT} - 1) \tag{8.17}$$

Similarly,

$$I_C = \frac{-I_{EC}}{\beta_R} - I_{EC} + I_{CC} = -I_{EC}\left(1 + \frac{1}{\beta_R}\right) + I_{CC}$$

$$= -I_{CS}(e^{qV_{BC}/kT} - 1) + \alpha_F I_{ES}(e^{qV_{BE}/kT} - 1) \tag{8.18}$$

The emitter and collector currents given in (8.17) and (8.18) are identical to those given in (8.5) and (8.6). Therefore the terminal currents for both the transport version and the injection version of the Ebers–Moll model are the same.

The transistor saturation current I_s changes with temperature and can be expressed as follows

$$I_s(T) = I_s(T_{nom})\left(\frac{T}{T_{nom}}\right)^3 \times \exp\left[-(E_g/k)\left(\frac{1}{T} - \frac{1}{T_{nom}}\right) \right] \tag{8.19}$$

where

T is the analysis temperature in Kelvin,
T_{nom} is the nominal temperature in Kelvin at which the transistor data is obtained

and

E_g is the energy gap of the semiconductor material of the transistor in ev.

A second-order mechanism requires a modification of this model. The mechanism, known as the base-width modulation, results from a voltage dependence of the width of the collector junction space-charge region. This causes the effective base width to be voltage dependent. We find the change in base-width, W, due to change in the base-collector voltage, V_{BC}, by using the Taylor series expansion of the equation $W = f(V_{BC})$ as follows,

$$W(V_{BC}) \simeq W(0) + V_{BC} \cdot \frac{dW}{dV_{BC}}\bigg|_{V_{BC}=0}$$

where $W(0)$ is the base-width at zero base-collector voltage, or,

$$\frac{W(V_{BC})}{W(0)} \simeq 1 + \frac{V_{BC}}{W(0)} \cdot \frac{dW}{dV_{BC}}\bigg|_{V_{BC}=0} \tag{8.20}$$

Again

$$\frac{kT}{q} \cdot \frac{1}{W(0)} \cdot \frac{dW}{dV_{BC}} = \eta \tag{8.21}$$

where η is the dimensionless base-width modulation factor. From (8.20) and (8.21) we may write

$$\frac{W(V_{BC})}{W(0)} \simeq 1 + V_{BC} \cdot \eta \cdot \frac{q}{kT} \simeq 1 + \frac{V_{BC}}{V_A}$$

or,

$$W(V_{BC}) \simeq W(0)\left(1 + \frac{V_{BC}}{V_A}\right) \tag{8.22}$$

V_A is known as the Early Voltage. It is proportional to the reciprocal of the rate of change in base-width.

The saturation current I_s and the common-emitter forward current gain β_F are both inversely proportional to the base-width and using (8.22) we can express them as follows,

$$I_s(V_{BC}) = I_s(0)/(1 + V_{BC}/V_A) \tag{8.23}$$

$$\beta_F(V_{BC}) = \beta_F(0)/(1 + V_{BC}/V_A) \tag{8.24}$$

The effect of base-width modulation is accounted for in the model by adding an output resistance r_0 and replacing the parameters I_s and β_F by (8.23) and (8.24) respectively. The output resistance r_0 is given by the ratio of the Early Voltage V_A to the collector current I_C,

$$r_0 = V_A/I_C \tag{8.25}$$

Resistances r'_b, r'_c and r'_e have been included in the transistor model to take into account the effects of bulk material in the base, collector and emitter respectively as shown in Fig. 8.4. These are dependent on the operating conditions, but for simplicity, we consider them constant.

Fig. 8.5. (a) Ebers–Moll model and (b) its simplified form in the forward active region.

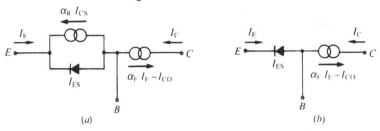

Simplified Ebers–Moll models

The regions of operation of a transistor are defined by the bias voltage on the two p–n junctions. Because each p–n junction may be either forward-biased or reverse-biased, four conditions of operation are possible. Simplified Ebers–Moll models can be derived for each of these regions. The n–p–n transistor model controlled by terminal currents as shown in Fig. 8.3 will be considered for this purpose.

The forward active region. The transistor is said to be operating in this region when the emitter junction is forward-biased and the collector junction is reverse-biased. In this case, if the collector junction voltage exceeds a few kT/q, the expression for the collector current given by (8.10) reduces to

$$I_C = -\alpha_F I_E + I_{CO} \tag{8.26}$$

The model shown in Fig. 8.5(a) represents the above equation together with the expression for the emitter current given by (8.5). The current source in parallel with the diode may be neglected since it is small compared to the typical value of the emitter current for this region. This will result in a model shown in Fig. 8.5(b).

The cutoff region. If both the emitter and collector junctions are reverse-biased, the transistor operates in the cutoff region. Examination of (8.5) and (8.6) shows that the terminal currents become independent of the voltages and remain constant when the voltages are in excess of a few kT/q. The Ebers–Moll model of the transistor for this case is drawn as shown in Fig. 8.6(a).

Fig. 8.6(a). Simplified Ebers–Moll model in cut-off region.

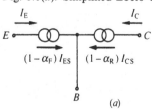

Fig. 8.6(*b*). Simplified Ebers–Moll model in saturation region.

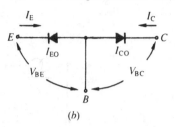

(*b*)

The saturation region. The transistor operates in the saturation region when both the emitter and collector junctions are forward-biased. The Ebers–Moll model for this region is the same as that shown in Fig. 8.3. If we neglect the current sources in parallel with the idealised diodes because of their comparatively small values, then the model is simplified to the form shown in Fig. 8.6(*b*).

The reverse active region. As the name implies, the transistor operates in this region when the junctions are biased in the opposite way to the one in the forward active region, i.e., the collector junction is forward-biased and the emitter junction is reverse-biased. The Ebers–Moll model for this case is shown in Fig. 8.7(*a*). Again if we ignore the current source in parallel with the idealised diode for the reason specified earlier the simplified model is as shown in Fig. 8.7(*b*). Designers most often avoid this operating region because the base-emitter junction cannot stand very high reverse-bias voltage (typically it can stand only -5 V). The area of the collector junction is greater than the area of the emitter junction to make the forward current gain α_F as large as possible, consequently making the reverse current gain α_R substantially lower than α_F.

The models we obtained above are nonlinear and the aid of a computer is essential for the analysis of a circuit. It is, however, possible to analyse a circuit even by hand computation if instead of using exponential $V–I$ characteristics for the diodes in the models shown in Figs. 8.5–8.7 we use

Fig. 8.7. (*a*) Ebers–Moll model and (*b*) its simplified form in reverse active region.

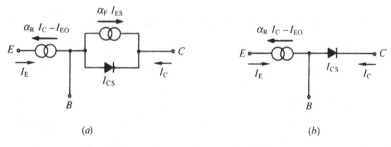

(*a*)   (*b*)

Fig. 8.8. Actual and piecewise linear V–I characteristic of a diode.

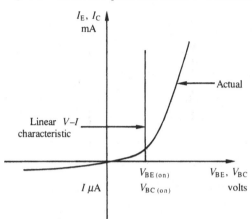

piecewise linear V–I characteristics as shown in Fig. 8.8. It can be observed that the diode current I_E (I_C) is greater than zero only when V_{BE} (V_{BC}) becomes equal to $V_{BE(ON)}$ $(V_{BC(ON)})$. Thus it is possible to replace a conducting diode by a battery having voltage equal to the forward voltage drop of the diode. The linear approximation Ebers–Moll models in all the regions of operation have been shown in Fig. 8.9. The current sources represent the reverse biased junctions. The following examples illustrate how these models can be used to obtain the quiescent values of voltages and currents.

Fig. 8.9. Linear approximation Ebers–Moll models in various regions of operation.

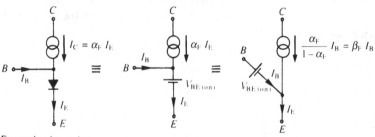

Fig. 8.9. (*cont.*)

Cutoff region

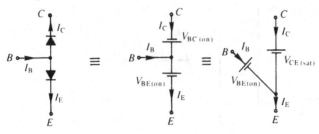

Saturated region

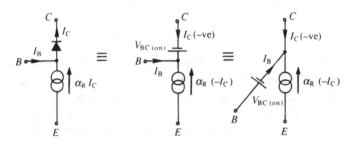

Reverse active region

Worked example 8.1

The transistor in the circuit of Fig. ex.8.1 has the following values of some parameters: $\beta_F = 100$, $V_{CE(SAT)} = 0.1$ V and $V_{BE(ON)} = 0.6$ V. Determine the values of the output voltages when the transistor is cutoff and saturated, and also the value of the input voltages at which the cutoff and saturation take place.

Fig. ex.8.1

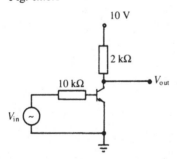

Solution

When the transistor is cutoff the circuit can be drawn as in Fig. ex.8.1(*a*):

Fig. ex.8.1(*a*)   Fig. ex.8.1(*b*)

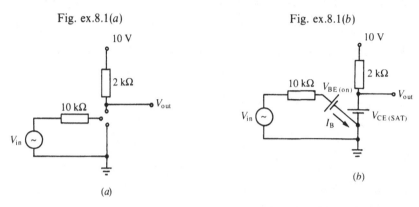

(*a*)

(*b*)

Since no current flows through the 2 kΩ resistor, there is no voltage drop across it and $V_{out} = 10$ volts.

When the transistor is saturated, the circuit can be drawn as shown in Fig. ex.8.1(*b*):

Now $V_{out} = V_{CE(SAT)} = 0.1$ V.

In order to find the input voltages at which the cutoff and saturation occur, let us consider that the transistor is in the forward active region. The circuit is then as shown in Fig. ex.8.1(*c*).

Fig. ex.8.1(*c*)

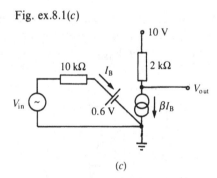

(*c*)

From the diagram we obtain

$$\frac{V_{in} - 0.6}{10 \times 10^3} = I_B \tag{1}$$

and also

$$V_{out} = 10 - 2 \times 10^3 \times \beta I_B \tag{2}$$

Since $\beta = 100$, from (1) and (2) we get

$$V_{out} = 22 - 20 V_{in} \tag{3}$$

When the transistor is cut off,

$$V_{out} = 10 \text{ volts}$$

Therefore from (3) we can write

$$10 = 22 - 20 V_{in}$$

or

$$V_{in} = 0.6 \text{ V}$$

When the transistor becomes saturated,

$$V_{out} = 0.1 \text{ V}$$

therefore from (3) we can write

$$0.1 = 22 - 20 V_{in}$$

or

$$V_{in} = 1.095 \text{ V}$$

Worked example 8.2

Plot and dimension the output waveform for the circuit shown in Fig. ex.8.2(a). The input is a symmetrical triangular waveform of ± 0.3 V magnitude having a frequency of 1 kHz. Given $V_{BE(ON)} = 0.2$ V and $\beta_F = 100$.

Fig. ex.8.2(a)

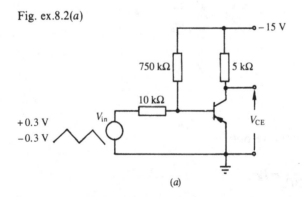

(a)

Fig. ex.8.2(b)

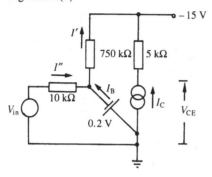

Solution

For instance suppose that the transistor is in the forward active region, then it can be replaced by Ebers–Moll model as shown in Fig. ex.8.2(b).

From the figure,

$$V_{CE} = -15 + 5 \times 10^3 \cdot I_C = -15 + 5 \times 10^3 \times 100 I_B$$

but,

$$I_B = I' - I''$$

$$= \frac{-0.2 - (-15)}{750 \times 10^3} - \frac{V_{in} - (-0.2)}{10 \times 10^3}$$

$$= \frac{14.8}{750 \times 10^3} - \frac{V_{in}}{10^4} - \frac{0.2}{10^4}$$

Therefore

$$V_{CE} = -15 + 5 \times 10^5 \left(\frac{14.8}{750 \times 10^3} - \frac{V_{in}}{10^4} - \frac{0.2}{10^4} \right)$$

$$= -15.13 - 50 V_{in} \tag{1}$$

From (1) we can say when $V_{in} = 0.3$ V, $V_{CE} = -30.13$ V, however as the plot shown in Fig. ex.8.2(c) shows, V_{CE} cannot be less than the negative power supply voltage. So $V_{CE} = -15$ V.

Again from (1), when

$$V_{in} = -0.3 \text{ V}, \quad V_{CE} = -0.13 \text{ V}$$

Fig. ex.8.2(c)

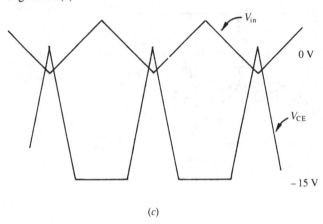

V_{in}

0 V

V_{CE}

−15 V

(c)

When the transistor is in the cutoff region,

$$V_{CE} = -15\,\text{V}$$

and from (1) we can say that

$$V_{in} = -\frac{0.13}{50} = -0.0026\,\text{V}$$

Dynamic behaviour. There are two types of charge storage in a transistor: the mobile charge in transit (current dependent) and the fixed charge in the depletion region (voltage dependent). The mobile charge has two components, one linked with the reference current at the collector, I_{CC} and the other linked with the reference current at the emitter I_{EC}. Their effects are included in the model by adding diffusion capacitances across each junction. They are given by

$$C_{DC} = Q_{DC}/V_{BC} = \tau_R I_{EC}/V_{BC}$$

and

$$C_{DE} = Q_{DE}/V_{BE} = \tau_F I_{CC}/V_{BE} \tag{8.27}$$

where

Q_{DC} and Q_{DE} are the mobile charges,

τ_F and τ_R are the forward and reverse transit times of the charges

The fixed charge can be modelled by junction capacitances which are approximately related to the junction voltages by the following expressions,

Fig. 8.10. Variation of junction capacitance with bias voltage.

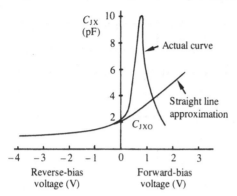

$C_{JX} = C_{JXO}/(1 - V_{BX}/\phi)^m \ for \ V_{BX} < \phi/2$

$\qquad = 2^m \cdot C_{JXO}[2m(V_{BX}/\phi) + (1 - m)] \ for \ V_{BX} > \phi/2 \qquad (8.28)$

where

$\qquad$ X is either E for the emitter or C for the collector junction,

$\qquad$ C_{JXO} is the zero-bias junction capacitance,

$\qquad$ m is the gradient factor of the junction

and

$\qquad$ ϕ is the barrier potential

Equation (8.28) shows that the junction capacitance is generally determined from zero-bias junction capacitance C_{JXO} and the barrier potential ϕ, and is exponentially dependent on the junction voltage V_{BX}. A typical plot of the variation of the junction capacitance with the bias voltage is shown in Fig. 8.10. The straight line approximation of the curve for $V_{BX} > \phi/2$ is valid in this case, since the diffusion capacitances are more prominent than the junction capacitances when the junctions are forward biased. The total capacitance across each junction is given by the sum of the respective diffusion and junction capacitances.

Apart from the diffusion and junction capacitances mentioned above, there is a substrate capacitance C_{sub} which is important in integrated circuits. Epitaxial layer substrate junctions are reverse biased for isolation purposes and according to the plot shown in Fig. 8.10 the substrate capacitance can be modelled as a constant value capacitor. Figure 8.11 shows the large-signal bipolar transistor model.

Fig. 8.11. Large-signal n–p–n transistor model.

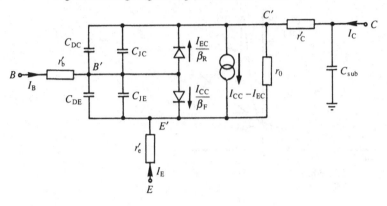

Fig. 8.12. Small-signal n–p–n transistor model.

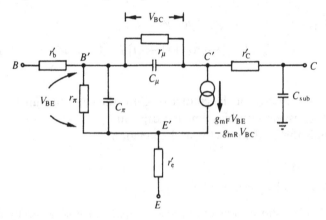

AC small-signal model. The small-signal circuit model of the transistor for a.c. analyses is shown in Fig. 8.12. The new elements in this model are governed by the following relations

$$r_\pi = \frac{\beta_F}{g_{mF}}, \; r_\mu = \frac{\beta_R}{g_{mR}} \tag{8.29}$$

$$C_\pi = C_{DE} + C_{JE}, \; C_\mu = C_{DC} + C_{JC} \tag{8.30}$$

where g_{mF} is the forward transconductance given by

$$\left.\frac{dI_{CC}}{dV_{BE}}\right|_{V_{BC=0}} = \frac{qI_{CC}}{kT}$$

and g_{mR} is the reverse transconductance given by

$$\left.\frac{dI_{EC}}{dV_{BC}}\right|_{V_{BE=0}} = \frac{qI_{EC}}{kT}$$

These are computed after d.c. operating points are determined. In the forward active region of operation, the reverse transconductance g_{mR} is essentially zero since I_{EC} is negligible. Thus resistance r_μ can be regarded as infinite and capacitance C_μ becomes nearly equal to C_{JC}. This leads to the well-used linear hybrid π model.

The parameters mentioned above for the three models completely describe the behaviour of a transistor in all the regions of operation and they can be determined from the measurements made at the device terminals as discussed in the following section.

The Ebers–Moll model can be used in several well-known computer aided design programs where the program may generate response equations involving nonlinear terms and solve by some iterative method, or, it may make a piecewise-linear approximation to the equations itself to simplify the response calculations. Some programs store the Ebers–Moll model as the prime model and needs certain parameters only to analyse transistor circuit problems as illustrated in the following example.

Worked example 8.3

Find the variations in current and voltage gains and also the input admittance of the amplifier circuit shown in Fig. ex.8.3 with frequency. The Ebers–Moll model of the transistor BCY70 is available as the prime model.

Fig. ex.8.3

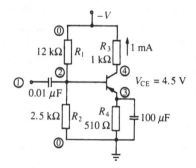

Solution

AUDIO AMPLIFIER. BASAK.

R1 12?3 0 2 ;
R2 2.5?3 2 0 ;
R3 ?3 0 4 ;
R4 510 3 0 ;
C1 0.01 ?-6 1 2 ;
C2 100?-6 3 0 ;

DB1 PNP 'MULLARD@ 'BCY70@ 4.5 1 ?-3 NODES 3 2 4 ;
                                V_{CE} I_C

PORTS 1 0 4 0 ;   ←   Input Output
OUTPUT(IG2, YI1, VG2) ;   ports ports
STEP LOG 100 10?4 5 ;
       ↑   ↑ ↑

REPEAT ; Hz
END ;   Start Step Step

FREQUENCY	INPUT ADMITTANCE PARALLEL FORM		VOLTAGE GAIN I/P-O/P		CURRENT GAIN I/P-O/P	
HZ	REAL	IMAG	MOD	PHASE	DB	PHASE
100.000	6.582?-08	6.275?-06	.3327	293.6	−65.51	24.16
158.489	1.605?-07	9.943?-06	.5604	285.0	−64.98	15.93
251.189	3.974?-07	1.575?-05	.9118	278.8	−64.75	10.22
398.107	9.911?-07	2.494?-05	1.460	274.6	−64.66	6.468
630.957	2.477?-06	3.943?-05	2.322	270.4	−64.62	4.040
1000.00	6.178?-06	6.212?-05	3.675	266.8	−64.60	2.467
1584.89	1.528?-05	9.698?-05	5.784	262.5	−64.60	1.421
2511.89	3.695?-05	1.482?-04	8.999	256.7	−64.59	.6817
3981.07	8.495?-05	2.154?-04	13.65	248.6	−64.59	.0890
6309.57	1.759?-04	2.831?-04	19.64	237.7	−64.59	359.5
10000.0	3.068?-04	3.156?-04	25.94	224.7	−64.59	358.8
15848.9	4.358?-04	2.925?-04	30.91	211.8	−64.60	357.9
25118.9	5.235?-04	2.399?-04	33.88	201.2	−64.61	356.5
39810.7	5.693?-04	1.960?-04	35.32	193.4	−64.63	354.4
63095.7	5.905?-04	1.797?-04	35.94	188.1	−64.70	351.1
100000	6.009?-04	1.980?-04	36.20	184.3	−64.85	346.1

END

Note: Here small signal analysis has been used to obtain the frequency response together with the input admittances.

8.1.2 Models for field effect transistors

Field effect transistors have three terminals like bipolar transistors; but detailed modelling requires inclusion of the substrate as a terminal. In this section we will consider only the metal-oxide semiconductor type FET. The characteristics of these FETs, especially the high circuit packing density and low power dissipation, have made the large scale integration possible.

The electrical characteristics of MOS transistors can be much more easily described and modelled in terms of the physical device parameters than those of the bipolar transistors. MOS devices do not show any charge storage effects comparable to bipolar transistors and the distributed gate-channel capacitance can be represented by a constant average value without much loss of accuracy. Also, due to the high impedance nature of these devices, the parasitic bulk series resistance associated with both the source and the drain terminals may be neglected.

The symbol and the model for an n-channel MOSFET is shown in Fig. 8.13. The model applies before saturation and also in the saturation region. Before saturation the effect of electric field, normal or parallel to the surface, on mobility is considered. For the saturation region the effect of mobile carriers on the drain-channel space-charge layer width is taken into account using an approximate analysis of two-dimensional effects. It can be shown, after making assumptions and approximations for the sake of simplicity, that when the drain-source voltage is less than the drain-source saturation voltage, that is before saturation, the current is given by

$$I_{DS} = A[V_G V_{DS} - V_{DS}^2/2] \tag{8.31}$$

Note: $A = \dfrac{\mu_0 C_{OX} Z}{L(1 + \theta V_G)}$; its unit is $A \cdot V^{-2}$, where A is the amplification

Fig. 8.13. The symbol and model of an n-channel MOSFET.

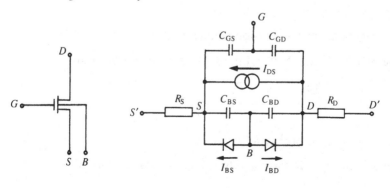

factor. It is a function of the carrier mobility μ_0, the thin-oxide capacitance per unit area C_{OX}, the channel width Z, the source-drain spacing L, the coefficient of the effect of electric field on mobility θ and the effective gate-source voltage V_G.

V_G, the effective gate-source voltage is the difference between V_{GS}, the gate-source voltage of the device and V_S, the threshold voltage. The threshold voltage lies between 0 and 1.5 volts and is a constant for a given technology, and V_{DS} is the drain-source voltage in volts.

At the pinchoff point, when the drain-source voltage becomes equal to the drain-source saturation voltage, the saturation current can be expressed as

$$I_{DSS} = A[V_G V_{DSS} - V_{DSS}^2/2] \tag{8.32}$$

where V_{DSS}, the drain-source saturation voltage is given by

$$V_{DSS} = V_G + v_c - [V_G^2 + v_c^2]^{1/2} \tag{8.33}$$

and v_c is a function of the electric field at pinchoff and the drain-source channel length. For large values of channel length, $v_c \gg V_G$, and from (8.32) and (8.33) we may write

$$I_{DSS} = A V_G^2/2 \tag{8.34}$$

The access resistances to the source and the drain, R_G and R_D can be obtained from the expressions:

$$R_S \simeq R W_S/2Z_S + R_C \tag{8.35}$$

$$R_D \simeq R W_D/2Z_D + R_C \tag{8.36}$$

where

R is the diffusion sheet resistivity in ohms
W_D and W_S are the diffusion widths in m
Z_S and Z_D are the channel widths in m

and

R_C is the metal-diffusion contact resistance

The bulk-drain and bulk-source diode currents are given by

$$I_{BD}' = I_{SD}'[\exp(q V_{BD}/kT) - 1] \tag{8.37}$$

and

$$I_{BS}' = I_{SS}'[\exp(q V_{BS}'/kT) - 1] \tag{8.38}$$

where I_{SD}' and I_{SS}' are the respective diode saturation currents.

The capacitances which depend mainly on the geometry and technology may be assumed to be independent of bias voltages and reference currents as mentioned earlier. Thus,

$$C = \frac{\varepsilon_0 \varepsilon_R a}{T_{OX}} = C_{OX} a \qquad (8.39)$$

where

>a is the area in m^2
>ε_0 is the permittivity of free space in Fm^{-1}
>ε_R is the dielectric constant
>T_{OX} is the oxide thickness in m

and

>C_{OX} is the thin-oxide capacitance per unit area

8.1.3. Macromodels

The idea and use of macromodels in designing electronic circuit are very common at the system level. They characterise the terminal behaviour of functional blocks such as operational amplifiers, gates, modulators, demodulators, etc. with sufficient accuracy.

A typical operational amplifier may contain twenty transistors in addition to resistors and capacitors, and use of the twelve-component Ebers–Moll model shown in Fig. 8.11 would lead to nearly three hundred components in the operational amplifier representation. Therefore analysis of a system with many operational amplifiers at the device level will create a very large analysis problem. On the other hand macromodels which provide adequate pin-to-pin representations of functional blocks may simplify the analysis of a system built with several functional blocks, typically by a factor of five.

A macromodel may be developed from the physical properties of the devices, measured data, manufacturers' specifications, or a combination thereof. In this section an operational amplifier is considered for developing a model since it is the most commonly used functional block in analogue integrated circuitry.

The circuit of an operational amplifier can be divided into three stages: input, intermediate and output. In the input stage the active load, the balance-to-unbalance converter and the circuitry employed for biasing can be replaced with ideal elements. A simple differential stage consisting of ideal transistors Q_1 and Q_2 can substitute for the composite transistors in

Fig. 8.14. Circuit diagram of the operational amplifier macromodel
(after G. R. Boyle *et al.* ©197-IEEE).

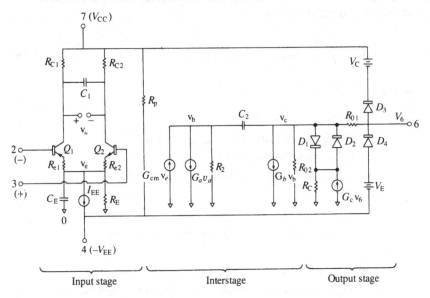

The input stage as shown in Fig. 8.14. Each of these transistors can now be
represented by the simplest Ebers–Moll model having two ideal p–n
junctions and two dependent current sources. Capacitors C_E and C_1 are
used to introduce second order effects for the slew rate and the phase
response respectively.

The differential-mode and common-mode voltage gains are provided by
the intermediate and output stage elements, G_{cm}, G_a, R_2, G_b and R_{02}. In
order to provide the necessary a.c. output resistance change with frequency,
a feedback capacitor C_2 is used in the model.

In the output stage, the elements D_1, D_2, R_c, G_c and R_{01} provide the
desired maximum short-circuit current. The elements D_3, D_4, V_C and V_E
comprise the output bounding circuits and determine the maximum output
voltage swing.

Design equations for the operational amplifier macromodel. The slew rate of
an operational amplifier is an important parameter in high-speed circuits
and can be used to determine the collector currents of the transistors Q_1 and
Q_2. For unity gain amplifiers Solomon *et al.* have shown that the positive
going slew rate $S^+ = 2I_{c1}/C_2$. Now if it is assumed that the quiescent
collector currents are equal for the transistors Q_1 and Q_2, then,

$$I_{c1} = I_{c2} = \tfrac{1}{2}C_2 S^+ \tag{8.40}$$

The charge-storage effects in the input stage makes the negative going slew rate S^- smaller than the positive going slew rate and is modelled by the capacitor C_E. From

$$S^- = 2I_{c1}/(C_2 + C_E)$$

we get

$$C_E = 2I_{c1}/S^- - C_2 \tag{8.41}$$

The current gains of the transistors Q_1 and Q_2 are given by the ratios of the collector currents to the base currents. The latter can be determined from the specifications for the average input bias current and the input offset current. The offset voltage of the operational amplifier can be modelled by specifying the saturation currents for the transistors.

Since

$$I_{c1} = I_{c2}, \; V_{os} = V_{BE1} - V_{BE2} = (kT/q)\ln(I_{s1}/I_{s2})$$

Hence

$$I_{s2} = I_{s1}\, e^{qV_{os}/kT} = I_{s1}[1 + qV_{os}/kT] \tag{8.42}$$

since

$$qV_{os}/kT \ll 1$$

The value of collector resistors in the input stage is obtained from f_T, the required unity gain bandwidth of the operational amplifier which is the product of the differential-mode voltage gain A_{DM} and the corner frequency of the frequency response f_0. By using a Miller-effect approximation in the intermediate stage the corner frequency can be found in terms of the circuit elements as follows,

$$f_0 = \frac{1}{2\pi R_2 C_2(1 + G_b R_{02})} \simeq \frac{1}{2\pi R_2 C_2 G_b R_{02}} \tag{8.43}$$

Again the low frequency differential-mode voltage gain is given by

$$A_{DM} = \frac{v_c}{v_b} \cdot \frac{v_b}{v_a} \cdot \frac{v_a}{v_{in}} \simeq G_a R_2 \cdot G_b R_{02} \cdot 1 = G_a G_b R_2 R_{02} \tag{8.44}$$

Therefore,

$$f_T = A_{DM} f_0 = G_a/2\pi C_2 \tag{8.45}$$

For convenience G_a may be chosen to be $1/R_{c1}$ and this yields

$$R_{c1} \text{ and } R_{c2} = 1/2\pi f_T C_2 \tag{8.46}$$

The resistors R_{E1} and R_{E2} are introduced in the model to simulate better the operational amplifiers which use emitter resistors to increase the slew rate. These can be found from the differential-mode voltage gain of the input stage

$$V_a/V_{in} = \frac{\beta_1 R_{c1} + \beta_2 R_{c2}}{\beta_1/g_{m1} + (\beta_1 + 1)R_{E1} + \beta_2/g_{m2} + (\beta_2 + 1)R_{E2}} \tag{8.47}$$

Since this stage has a gain of unity and $I_{c1} = I_{c2}$, we may write

$$R_{E1} = R_{E2} = \frac{\beta_1 + \beta_2}{\beta_1 + \beta_2 + 2}\left(R_{c1} - \frac{1}{g_{m1}}\right) \tag{8.48}$$

Since $I_{c1} = I_{c2}$

$$g_{m1} = g_{m2}$$

and the value of the d.c. current source in the input stage is given by

$$I_{EE} = \left(\frac{\beta_1 + 1}{\beta_1} + \frac{\beta_2 + 1}{\beta_2}\right)I_{c1} \tag{8.49}$$

The resistance R_E is the output resistance of the transistor which acts as the current source. It is the ratio of the 'early voltage' V_A of the transistor to its collector current which is approximately equal to I_{EE}.

The capacitor C_1 introduces excess phase effects in the differential-mode amplifier response by generating another pole at

$$f_2 = -1/2\pi 2R_{c1}C_1 \tag{8.50}$$

The excess phase at unity gain bandwidth due to this pole can then be found from (8.46) and (8.50) to be

$$\Delta\phi = \tan^{-1}\frac{f_T}{|f_2|} = \tan^{-1}\frac{2C_1}{C_2} \tag{8.51}$$

from which we get $C_1 = (C_2/2)\tan\Delta\phi$.

The common-mode voltage gain in the input stage from v_{in} to v_e is approximately unity since R_E is large (of the order of one hundred kΩ). The common-mode voltage gain from the input to v_b is then

$$v_{bcm}/v_{incm} = v_{bcm}/v_e \simeq G_{cm}R_2$$

and the differential-mode voltage gain from the input to v_b is

$$v_{bdm}/v_{indm} = G_a R_2 = R_2/R_{c1}$$

We chose $G_a = \dfrac{1}{R_{C1}}$ for convenience earlier ((8.45) and 8.46)).

Since the common-mode rejection ratio is the ratio of the above two gains we may write

$$G_{cm} = 1/(CMRR)R_{c1} \qquad (8.52)$$

Thus by choosing the appropriate value for G_{cm} we can model the CMRR of the operational amplifier.

The output resistance R_{out} of the model at very low frequencies for the quiescent state is given by $R_{01} + R_{02}$. However at high frequencies, R_{02} is shorted out by the capacitor C_2 and the output resistance becomes equal to R_{01}. Thus R_{01} models the a.c. output resistance of the operational amplifier R_0. The difference between the d.c. output resistance R_{out} and the a.c. output resistance gives the value of R_{02}.

By substituting an appropriate value for R_2 in (8.44) G_b can be computed,

$$G_b = A_{DM}R_{c1}/R_2R_{02} \qquad (8.53)$$

The output voltage V_6 appears across R_C, since the $R_C - G_C V_6$ combination is an equivalent to a voltage-controlled source. When both the voltage-clamping diodes D_3 and D_4 are off, the maximum output current is

$$I_{0(max)} = V_D/R_{01} \qquad (8.54)$$

where

$$V_D = kT/q \ln \frac{I_{D(max)}}{I_{SD}} \qquad (8.55)$$

$I_{D(max)}$ is the maximum current through D_1 and D_2 and I_{SD} is the saturation current of diodes D_1 and D_2.

The maximum voltage drop across R_{01} is $I_{0(max)}R_{01}$ which is also the maximum voltage drop across D_1. Therefore,

$$I_{D(max)} = I_{SD} \exp[qI_{0(max)}R_{01}/kT] \qquad (8.56)$$

Again it can be observed from Fig. 8.14 that the maximum current available from the $G_b v_b$ is

$$I_{max} = I_{D(max)} + I_{0(max)} = G_b v_b$$
$$= G_a G_b v_a R_2 = G_b v_a R_2/R_{c1} = 2I_{c1}R_2 G_b \qquad (8.57)$$

From (8.56) we obtain

$$I_{SD1} = I_{SD2} = I_{D(max)} \exp\left[\frac{-qI_{0(max)}R_{01}}{kT}\right] \qquad (8.58)$$

We must make R_c very small so that we have a good approximation to the voltage-controlled voltage source. If we assume that the voltage across R_C is only one percent of V_{D1} or V_{D2}, then it can be shown that

$$R_c = \frac{kT}{100qI_{D(max)}} \ln(I_{D(max)}/I_{SD1})$$ (8.59)

The required value for the coefficient of the voltage controlled current source $G_c v_6$ is then $1/R_c$.

The d.c. voltage sources V_C and V_E and the diodes D_3 and D_4 limit the output voltage swing. When the output is large and positive the diode D_3 conducts and the output voltage is

$$V_{out}{}^+ = V_{CC} - V_C + V_{D3} = V_{CC} - V_C + \frac{kT}{q} \ln \frac{I_{0(max)}{}^+}{I_{SD3}}$$ (8.60)

Therefore the value of the positive d.c. voltage source is

$$V_C = V_{CC} - V_{out}{}^+ + \frac{kT}{q} \ln [I_{0(max)}{}^+/I_{SD3}]$$ (8.61)

Similarly,

$$V_E = V_{EE} - V_{out}{}^- + \frac{kT}{q} \ln [I_{0(max)}{}^-/I_{SD4}]$$ (8.62)

Resistor R_p is introduced to model the actual d.c. power dissipation P_d in the operational amplifier. From Fig. 8.14 we may write

$$P_d = V_{CC} \cdot 2I_{c1} + V_{EE} \cdot I_{EE} + (V_{CC} + V_{EE})^2/R_p$$

or,

$$R_p = \frac{(V_{cc} + V_{EE})^2}{P_d - V_{CC} \cdot 2I_{c1} - V_{EE}I_{EE}}$$ (8.63)

The design equations for the parameters and element values of a macromodel of an operational amplifier have been derived above, The model accurately represents the circuit behaviour for non-linear d.c., small signal a.c. and large signal transient responses.

Worked example 8.4

Develop a macromodel for the operational amplifier LM118 using the data given in Table ex.8.4.

Table ex.8.4. *Data of LM118*

C_2	5 pF	R_0	32 Ω
$S_R{}^+$	100 V/μs	$I_{0(max)}{}^+$	25 mA
$S_R{}^-$	71 V/μs	$I_{0(max)}{}^-$	25 mA
I_B	120 nA	V^+	13 V
I_{os}	6 nA	V^-	−13 V
V_{os}	2 mV	P_d	80 mW
$\Delta\phi$	40°	A_{DM}	16×10^3
CMMR	100 dB	at 1 kHz	
R_{out}	75 Ω	A_{DM}	2×10^5

Solution

Let us consider the model shown in Fig. 8.14. Collector currents I_{c1} and I_{c2} are found by considering the positive going slew rate. From (8.40)

$$I_{c1} = \tfrac{1}{2}C_2S_R{}^+ = \tfrac{1}{2} \times 5 \times 10^{-12} \times 100 \times 10^6 = 250\,\mu A$$

Charge storage effects can be modelled by C_E. Using (8.41)

$$C_E = \frac{2I_{c1}}{S_R{}^-} - C_2 = \frac{2 \times 250 \times 10^{-6}}{71 \times 10^6} - 5 \times 10^{-12} = 2.04\,pF$$

Base currents

$$I_{B1} = I_B + \frac{I_{os}}{2} = 120 + 3 = 123\,nA$$

$$I_{B2} = I_B - \frac{I_{os}}{2} = 120 - 3 = 117\,nA$$

Therefore the current gains are

$$\beta_1 = \frac{250 \times 10^{-6}}{123 \times 10^{-9}} = 2030 \text{ and } \beta_2 = \frac{250 \times 10^{-6}}{117 \times 10^{-9}} = 2137$$

Now from (8.49)

$$I_{EE} = \left(\frac{2030+1}{2030} + \frac{2137+1}{2137}\right)250 \times 10^{-6}$$

$$= 500\,\mu A$$

Therefore the output resistance of the current source,

$$R_E = \frac{200}{500 \times 10^{-6}} = 400\,k\Omega \quad \text{assuming } V_A = 200 \text{ volts}$$

For a small n–p–n IC transistor $I_{s1} = 5 \times 10^{-16}$ A, say, then from (8.42)

$$I_{s2} = 5 \times 10^{-16}(1 + \tfrac{2}{26}) = 5.38 \times 10^{-16} \text{ A}$$

The unity gain bandwidth is obtained from (8.45)

$$f_T = A_{DM} f_0 = 16 \times 10^3 \times 1 \times 10^3 = 16 \text{ MHz}$$

For convenience choose

$$G_a = \frac{1}{R_{c1}}$$

Therefore from (8.46)

$$R_{c1} = R_{c2} = \frac{1}{2\pi f_T C_2}$$

$$= \frac{1}{2\pi \times 16 \times 10^6 \times 5 \times 10^{-12}}$$

$$= 1.99 \text{ k}\Omega$$

The transconductance $g_{m1} = \dfrac{qI_C}{kT} = \dfrac{250 \times 10^{-6}}{26 \times 10^{-3}} = 9.62 \text{ m}\mho$. From (8.48)

$$R_{E1} = R_{E2} = \frac{\beta_1 + \beta_2}{\beta_1 + \beta_2 + 2} \left[R_{c1} - \frac{1}{g_{m1}} \right]$$

$$= \frac{2030 + 2137}{2030 + 2137 + 2}(1990 - 104)$$

$$= 1.89 \text{ k}\Omega$$

Using (8.51) to model the excess phase shift

$$C_1 = \frac{C_2}{2} \tan 40° = \frac{5 \times 10^{-12}}{2} \times 0.8391 = 2.1 \text{ pF}$$

To model the CMMR of the operational amplifier we use (8.52)

$$G_{cm} = \frac{1}{(CMRR)R_{C1}} = \frac{1}{10^5 \times 1.99 \times 10^3} = 5.02 \text{ n}\mho$$

Given $R_{01} = R_0 = 32 \Omega$. Therefore

$$R_{02} = R_{out} - R_{01} = 43 \Omega$$

From (8.53)

$$G_b = \frac{A_{DM}R_{c1}}{R_2 R_{02}} = \frac{2 \times 10^5 \times 1990}{100 \times 10^3 \times 43} = 92.5 \mho$$

(For the interstage we assumed $R_2 = 100\,\text{k}\Omega$.) Now from (8.57)

$$I_{D(max)} = 2I_{c1}R_2G_b - I_{0(max)}$$

$$= 2 \times 250 \times 10^{-6} \times 100 \times 10^3 \times 92.5 - 25 \times 10^{-3}$$

$$= 4624.9\,\text{A}$$

(8.58) gives

$$I_{SD1} = I_{D(max)}e^{-qI_{0(max)}R_{01}/kT}$$

$$= 4624.9\,e^{\dfrac{-25 \times 10^{-3} \times 32}{26 \times 10^{-3}}}$$

$$= 200.5\,\text{pA}$$

(8.59) gives

$$R_C = \frac{kT}{100qI_{D(max)}}\ln(I_{D(max)}/I_{SD1})$$

$$= \frac{26 \times 10^{-3}}{100 \times 4624.1}\ln\frac{4624.1}{200.5 \times 10^{-12}}$$

$$= 1.73\,\text{m}\Omega$$

Therefore the value of

$$G_c = \frac{1}{R_C} = 578\,\mho$$

Now from (8.61) and (8.62)

$$V_C = 15 - 13 + 26 \times 10^{-3}\ln\frac{25 \times 10^{-3}}{5 \times 10^{-16}} = 2.82\,\text{volts}$$

$$V_E = 15 - 13 + 26 \times 10^{-3}\ln\frac{25 \times 10^{-3}}{5 \times 10^{-16}} = 2.82\,\text{volts}$$

Lastly (8.63) gives

$$R_p = \frac{(V_{CC} + V_{EE})^2}{P_d - V_{CC}2I_C - V_{EE}I_{EE}}$$

$$= \frac{(15 + 15)^2}{80 \times 10^{-3} - 15 \times 2 \times 250 \times 10^{-6} - 15 \times 500 \times 10^{-6}}$$

$$= 13.8\,\text{k}\Omega$$

8.2.   Analysis of circuits

The initial evaluation or testing of a circuit is essentially carried out to ensure that it meets the specification. From the computational point of view, the breadboarding and circuit testing stages may be replaced by the analysis of the circuit using the models mentioned in the preceding sections. The circuit analysis techniques which are employed fall mainly in three categories: (a) d.c. analysis, (b) a.c. small signal analysis and (c) transient analysis.

8.2.1   D.C. analysis

D.C. analysis is concerned with the computation of the bias conditions of a circuit. It is accomplished by the solution of a set of nonlinear circuit equations based on the algebraic characteristic of the circuit elements. The solution is obtained using iterative numerical techniques.

As mentioned earlier, the first step in the analysis of nonlinear networks is to determine the d.c. operating point which is then used to compute linearised model parameters for the a.c. small-signal analysis and initial conditions for the transient analysis. For d.c. analysis, all the capacitors are removed from the circuit and each nonlinear element is linearised about a presumed operating point. Then we may solve the linearised circuit for fixed values of d.c. sources to produce a new operating point with the help of the well known Newton–Raphson algorithm. The parameters of each transistor and diode are updated and the process is repeated until the node voltage on successive iterations converge to within specified tolerance. This widely used algorithm has fast convergence (quadratic if the initial estimate is close to the solution). The theoretical background is available in books on numerical analysis.

Let us now consider an example of a simple common-emitter amplifier circuit as shown in Fig. 8.15(a), to see how we can derive a set of equations for a circuit. The amplifier circuit may be modified to that shown in Fig.

Fig. 8.15. (a) Simple and (b) modified common-emitter amplifier circuit.

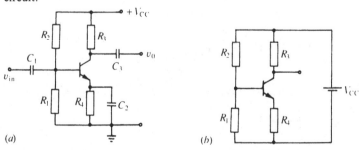

(a)   (b)

Fig. 8.16. Model of the common emitter amplifier circuit.

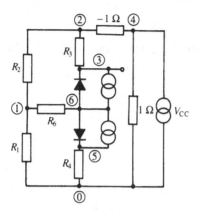

8.15(b) since the capacitors will behave as an open circuit for d.c. currents and voltages. The voltage source gives rise to difficulties in formulating equations and the most straightforward way of dealing with this is to convert the voltage source to a current source by using Norton's theorem. Figure 8.16 shows how the voltage source has been converted to current source by introducing two resistances of equal but opposite values in the circuit. This produces an extra node but simplifies the analysis. The n–p–n transistor has been replaced by the Ebers–Moll model. The resistance R_6 is the base-spreading resistance of the transistor. There are six nodes in the circuit as can be seen in the figure and we may write six nodal equations as follows:

$$\left.\begin{aligned}
&\frac{V_1}{R_1} + \frac{V_1 - V_2}{R_2} + \frac{V_1 - V_6}{R_6} = 0 \\[2mm]
&\frac{V_2 - V_1}{R_2} + \frac{V_2 - V_3}{R_3} + \frac{V_2 - V_4}{-1} = 0 \\[2mm]
&\frac{V_3 - V_2}{R_3} + I_{CS}(e^{q(V_6 - V_3)/kT} - 1) - \alpha_F I_{ES}(e^{q(V_6 - V_5)/kT} - 1) = 0 \\[2mm]
&V_4 + \frac{V_4 - V_2}{-1} = V_{CC} \\[2mm]
&\frac{V_5}{R_4} + I_{ES}(e^{q(V_6 - V_5)/kT} - 1) - \alpha_R I_{CS}(e^{q(V_6 - V_3)/kT} - 1) = 0 \\[2mm]
&\frac{V_6 - V_1}{R_6} - I_{CS}(e^{q(V_6 - V_3)/kT} - 1) + \alpha_R I_{CS}(e^{q(V_6 - V_3)/kT} - 1) \\[2mm]
&\qquad\qquad - I_{ES}(e^{q(V_6 - V_5)/kT} - 1) + \alpha_F I_{ES}(e^{q(V_6 - V_5)/kT} - 1) = 0
\end{aligned}\right\} \quad (8.64)$$

Equations (8.64) are in the form

$$I = Y(V) \qquad (8.65)$$

where $Y(V)$ is a matrix whose entries are functions of $V_1 \ldots V_6$. In this example

$$I = \begin{bmatrix} 0 \\ 0 \\ 0 \\ I_0 \\ 0 \\ 0 \end{bmatrix}$$

where I_0 has the value of V_{CC}.

Equation (8.65) may be written in the following form

$$Y(V) - I = 0 \qquad (8.66)$$

or in more concise form

$$F(V) = 0 \qquad (8.67)$$

The Newton–Raphson technique may be used to solve systems of equations like these.

Newton–Raphson method

Let us now consider for simplicity two non-linear equations f_1 and f_2 in two variables x_1 and x_2.

$$f_1(x_1, x_2) = 0$$
$$f_2(x_1, x_2) = 0 \qquad (8.68)$$

The equations (8.68) are in matrix form

$$F(X) = 0 \qquad (8.69)$$

where x_1 and x_2 are the components of matrix X. Assume that the set of equations has a solution which we denote by X^*. Now expanding (8.68) in terms of $(X^* - X)$ by Taylor's theorem, and neglecting the second-order terms (if X^* is close to X, then we may neglect the higher order terms)

$$f_1(X^*) = f_1(X) + \frac{\partial f_1}{\partial x_1}(x_1^* - x_1) + \frac{\partial f_1}{\partial x_2}(x_2^* - x_2)$$

and

$$f_2(X^*)=f_2(X)+\frac{\partial f_2}{\partial x_1}(x_1{}^*-x_1)+\frac{\partial f_2}{\partial x_2}(x_2{}^*-x_2) \tag{8.70}$$

Equating expressions (8.70) to zero and writing in matrix form

$$-\begin{bmatrix} f_1(X) \\ f_2(X) \end{bmatrix}=\begin{bmatrix} \dfrac{\partial f_1}{\partial x_1} & \dfrac{\partial f_1}{\partial x_2} \\ \dfrac{\partial f_2}{\partial x_1} & \dfrac{\partial f_2}{\partial x_2} \end{bmatrix}\begin{bmatrix} x_1{}^*-x_1 \\ x_2{}^*-x_2 \end{bmatrix}$$

or

$$-[F(X)]=[J][X^*-X]$$

where $[J]$ is the Jacobian matrix of the set of equations. Now if we say that an estimate of a solution is obtained after k iterations then we can indicate the iteration sequence by writing ks as superscripts as follows.

$$[X^{k+1}-X^k]=-[J]^{-1}[F(X^k)]$$

or,

$$X^{k+1}=X^k-[J]^{-1}[F(X^k)] \tag{8.71}$$

(8.71) is a special fixed-point algorithm for solving a set of equations $F(X)=0$. It is called the Newton–Raphson algorithm.

Worked example 8.5

In order to determine the d.c. conditions of the circuit of Fig. ex.8.5(a) obtain a set of general equations in matrix form which can be solved using the Newton–Raphson algorithm. Compute the d.c. voltages at the nodes of the circuit if the transistor is in forward active region. The

Fig. ex.8.5(a)

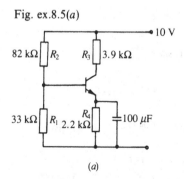

(a)

saturation currents of the diodes are 0.1 pA, the base-spreading resistance is $200\,\Omega$ and the common-base forward and reverse gains are 0.98 and 0.05 respectively. An initial guess may be obtained by assuming $V_{BE(ON)} = 0.7\,V$ and no base current is flowing.

Solution

After replacing the transistor by Ebers–Moll model (Fig. ex.8.5(b)), R_6 is the base-spreading resistance.

Fig. ex.8.5(b)

(b)

at node 1 $\dfrac{V_1}{R_1} + \dfrac{V_1 - V_2}{R_2} + \dfrac{V_1 - V_6}{R_6} = 0$

at node 2 $\dfrac{V_2 - V_1}{R_2} + \dfrac{V_2 - V_3}{R_3} + \dfrac{V_2 - V_4}{-1} = 0$

at node 3 $\dfrac{V_3 - V_2}{R_3} + I_{CS}(e^{(V_6 - V_3)q/kT} - 1)$

$\qquad\qquad - \alpha_F I_{ES}(e^{(V_6 - V_5)q/kT}) - 1) = 0$

at node 4 $\dfrac{V_4 - V_2}{-1} + \dfrac{V_4}{1} = V_{CC}$

at node 5 $\dfrac{V_5}{R_4} + I_{ES}(e^{(V_6 - V_5)q/kT} - 1)$

$\qquad\qquad - \alpha_R I_{CS}(e^{(V_6 - V_3)q/kT}) - 1) = 0$

at node 6 $\dfrac{V_6 - V_1}{R_6} - I_{CS}(e^{(V_6 - V_3)q/kT} - 1) + \alpha_R I_{CS}(e^{(V_6 - V_3)q/kT}) - 1)$

$\qquad\qquad - I_{ES}(e^{(V_6 - V_5)q/kT}$

$\qquad\qquad - 1) + \alpha_F I_{ES}(e^{(V_6 - V_5)q/kT}) - 1) = 0$

Since the transistor is in the forward active region, we may remove the current source $\alpha_R I_C$ and also replace the reverse biased base-collector diode by a resistor of say 22 kΩ or convenience. Then the circuit can be redrawn as shown in Fig. ex.8.5(c):

Fig. ex.8.5(c)

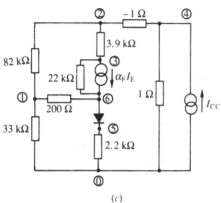

(c)

Initial values of all the node potentials are determined by assuming $V_{BE} = 0.7$ and $V_{16} = 0$ V. With the help of the following program; node potentials have been computed as shown in the print out.

```
10  REM SOLVING NODE VOLTAGE USING JACOBIAN MATRIX
    [dX] = INV[A]*-[F]
20  DIM F(6,1),A(6,6),V(6,1),C(6,6),D(6,1)
30  PRINT"ENTER RESISTANCE VALUE R1,R2,R3,R4,R6,R7"
40  INPUT R1,R2,R3,R4,R6,R7
50  K1 = (1/R1) + (1/R2) + (1/R6)
60  K2 = (1/R2) + (1/R3) - 1
70  PRINT"ENTER       INITIAL       NODE       VOLTAGES
    V1,V2,V3,V4,V5,V6"
80  INPUT V1,V2,V3,V4,V5,V6
90  PRINT"ENTER SUPPLY VOLTAGE Vcc"
100 INPUT V7
110 PRINT"ENTER NUMBER OF ITERATION"
120 INPUT I
130 E = 1E-13
140 G = 0.98
150 FOR S = 1 to I
160 GOSUB 500
170 MAT C = ZER(6.6)
```

```
180 MAT D=ZER(6,1)
190 MAT C=INV(A)
200 MAT D=C*F
210 V1=V1-D(1,1)
212 V2=V2-D(2,1)
214 V3=V3-D(3,1)
216 V4=V4-D(4,1)
218 V5=V5-D(5,1)
220 V6=V6-D(6,1)
230 PRINT"ITERATION = ";S
240 PRINT"V1= ";V1,"V2= ";V2
250 PRINT"V3= ";V3,"V4= ";V4
260 PRINT"V5= ";V5,"V6= ";V6
270 PRINT"V1error= ";D(1,1),"V2error= ";D(2,1)
280 PRINT"V3error= ";D(3,1),"V4error= ";D(4,1)
290 PRINT"V5error= ";D(5,1),"V6error= ";D(6,1)
300 NEXT S
310 STOP
500 REM FORMING -[F] MATRIX
510 F(1,1)=(V1*K1)-(V2/R2)-(V6/R6)
520 F(2,1)=(-V1/R2)+(K2*V2)-(V3/R3)+V4
530 F(3,1)=(−V2/R3)+(((1/R3)+(1/R7))*V3)−(V6/R7)−(G*E*((EXP
    ((V6−V5)/.026))−1))
540 F(4,1)=(−V4)+V4+V2−V7
550 F(5,1)=(V5/R4)+(E*((EXP((V6−V5)/.026))−1))
560 F(6,1)=(−V1/R6)−(V3/R7)+(((1/R6)+(1/R7))*V6)+(E*(G−1)*
    ((EXP((V6−V5)/.026))−1))
590 REM FORMING MATRIX [A]
600 A(1,1)=K1
610 A(1,2)=1/R2
620 A(1,3)=0
630 A(1,4)=0
640 A(1,5)=0
650 A(1,6)=1/R6
670 A(2,1)=−1/R2
680 A(2,2)=K2
690 A(2,3)=−1/R3
700 A(2,4)=1
710 A(2,5)=0
720 A(2,6)=0
```

```
730  A(3,1)=0
740  A(3,2)= −1/R3
750  A(3,3)=(1/R3)+(1/R7)
760  A(3,4)=0
770  A(3,5)=G∗E∗38.4615∗(EXP((V6−V5)/.026))
780  A(3,6)=(−G∗E∗38.4615∗(EXP((V6−V5)/.026)))−(1/R7)
790  A(4,1)=0
800  A(4,2)=1
810  A(4,3)=0
820  A(4,4)=0
830  A(4,5)=0
840  A(4,6)=0
850  A(5,1)=0
860  A(5,2)=0
870  A(5,3)=0
880  A(5,4)=0
890  A(5,5)=(1/R4)−(E∗38.4615∗(EXP((V6−V5)/.026)))
900  A(5,6)=E∗38.4615∗(EXP((V6−V5)/.026))
910  A(6,1)= −1/R6
920  A(6,2)=0
930  A(6,3)= −1/R7
940  A(6,4)=0
950  A(6,5)= −E∗(G−1)∗38.4615∗(EXP((V6−V5)/.026))
960  A(6,6)=(1/R6)+(1/R7)+(E∗(G−1)∗38.4615∗
     (EXP((V6−V5)/.026)))
970  RETURN
980  END


basic jacob.basic
ENTER RESISTANCE VALUE R1,R2,R3,R4,R6,R7
? 33e3,82e3,3.9e3,2.2e3,200,22e3
ENTER INITIAL NODE VOLTAGES V1,V2,V3,V4,V5,V6
? 2.8,10,5.5,10,2,2.7
ENTER SUPPLY VOLTAGE Vcc
? 10
ENTER NUMBER OF ITERATION
? 5
ITERATION= 1
V1= 2.68817      V2= 10
V3= 5.89262      V4= 9.99886
```

V5 = 2.03986   V6 = 2.71337
V1error =   0.111826   ` 2error =   0
V3error = −0.392624   V4error =   1.14225 E-3
V5error = −0.039857   V6error = −1.33677 E-2
ITERATION = 2
V1 = 2.70184   V2 = 10
V3 = 5.83505   V4 = 9.99884
V5 = 2.08019   V6 = 2.72632
V1error = −1.36668 E-2   V2error =   0
V3error =   0.057575   V4error =   1.45252 E-5
V5error = −4.03346 E-2   V6error = −1.29524 E-2
ITERATION = 3
V1 = 2.71511   V2 = 10
V3 = 5.77412   V4 = 9.99883
V5 = 2.12276   V6 = 2.73884
V1error = −1.32685 E-2   V2error =   0
V3error =   6.09304 E-2   V4error =   1.54907 E-5
V5error = −4.25669 E-2   V6error = −1.25234 E-2
ITERATION = 4
V1 = 2.72801   V2 = 10
V3 = 5.7004   V4 = 9.99881
V5 = 2.17394   V6 = 2.75088
V1error = −1.29049 E-2   V2error =   0
V3error =   7.37176 E-2   V4error =   1.87807 E-5
V5error = −5.11828 E-2   V6error = −1.20322 E-2
ITERATION = 5
V1 = 2.74079   V2 = 10
V3 = 5.55575   V4 = 9.99877
V5 = 2.2731   V6 = 2.76206
V1error = −1.27764 E-2   V2error =   0
V3error =   0.144648   V4error =   3.67981 E-5
V5error = −9.91624 E-2   V6error = −1.11795 E-2

8.2.2   A.C. small-signal analysis

We may use a.c. small-signal analysis to evaluate the performance of a circuit under various a.c. excitations. In this analysis, the non-linear elements are replaced by linearised equivalents which are computed after d.c. analysis. The voltage sources are short-circuited, the current sources in transistor models are open-circuited and the capacitors are replaced by complex frequency-dependent conductances

$$Y_c = j\omega C$$

Due to short-circuiting of voltage sources the number of nodes decreases and they have to be renumbered to form a new complex Y-matrix. This matrix is solved at sequential frequency points by the 'most efficient' all-round method, the Gaussian elimination algorithm. This method is based on the extremely elementary idea of eliminating the variables one at a time until there is only one equation in one variable left. This equation is then solved to give the solution for this one variable, say x_n. The value of x_n is then substituted back into the preceding equation to obtain the remaining solutions.

8.2.3 Transient analysis

The above two analysis techniques provide sufficient detail of the performance for many circuits; but in some cases where large-signal excursions are experienced, transient analysis is applied. For this, nonlinear time-domain differential equations which describe the circuit behaviour, are solved. The solution for the circuit is obtained at successive time intervals. The initial conditions are obtained from the d.c. analysis. The source excitation vector, I is now a function of time. The current through the capacitor is given by

$$i_c(t) = C \cdot dV/dt$$

where V is the capacitor branch voltage. The linear capacitor may be replaced as shown in Fig. 8.17 by an equivalent conductance,

$$G = 2C/t$$

in parallel with two current sources,

$$2C/t \cdot V^{n-1} \text{ and } i^{n-1}$$

where t is the time interval, V^{n-1} is the capacitor branch voltage and i^{n-1} is current through it at the previous time interval. The diode capacitance is assumed to be linear at a given time interval and its value is calculated by using the voltage across the diode and the current through it at the preceding time interval.

Fig. 8.17. Model of a linear capacitor (after N. D. Arora *et al.*).

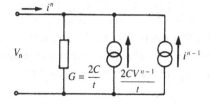

Thus the differential equations at the source are replaced by algebraic equations and the circuit becomes resistive. So the Y-matrix is real and formed in the same way as for d.c. analysis. In order to obtain good results it is important that accurate time intervals are used to solve the Y-matrix starting from a given time.

8.3 Programs

Electronic circuit simulator programs can perform d.c., a.c. and transient analysis. In the process of circuit designs these programs are used to perform repeated analysis of the circuit. These include the equation compilers, which generate the necessary equations from the node and element description of the circuit. The circuit description input allows considerable freedom in the description of the numerical values, the order of inputs, and the numbering of nodes. Some examples of these type of programs are SCEPTRE and ECAP. The general features of each are similar, and the reasons for choosing one program rather than the other are normally based on individual analysis needs and computer facilities.

Electronic Circuit Analysis Program (ECAP), which was developed by IBM and the Norden Division of United Aircraft uses the piecewise linear approach for the analysis of nonlinear circuits. It is an integrated analysis system consisting of four related programs. These are the input language program, the d.c. analysis program, the a.c. analysis program and the transient analysis program.

The first program is the communication link between the user and the other three programs. With the help of input statements made in simple formats, the circuit arrangement of components can be described, and the type of analysis as well as the desired output can be chosen. An important feature of an input language program is the use of an English-language style and a simple format for the input. Six input statements are used to define the circuit topology, the circuit element values, the type of analysis described, the circuit excitation, and the desired outputs in ECAP. The three other progams allow one to obtain respective solutions for the circuit.

Figure 8.18 shows the typical branch configurations. A circuit does not have to be restricted to a single dependent current source, and may have more than one in the same branch.

The initial step of the circuit analysis using ECAP is the preparation of a preliminary circuit schematic that may perform a desired function. The analysis ensures that the circuit will perform as expected. Next the physical components are replaced by their models. The nodes are then identified by circled numbers and the branch numbers are enclosed in squares. The

Fig. 8.18. ECAP branch configuration: (a) d.c. analysis, (b) a.c. analysis, (c) transient analysis (after M. F. Hordeski).

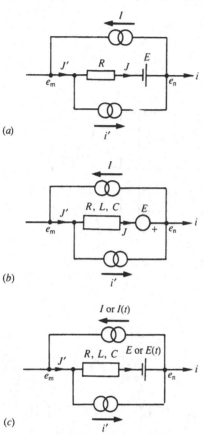

(a)

(b)

(c)

The types of circuit elements used in the programs are summarised below:

Circuit Element	Analysis		
	DC	AC	Transient
Resistor	x	x	x
Capacitor		x	x
Inductor		x	x
Mutual inductor		x	
Voltage source, fixed	x	x	x
Voltage source, time-dependent			x
Current source, fixed	x	x	x
Current source, time-dependent			x
Switch			x

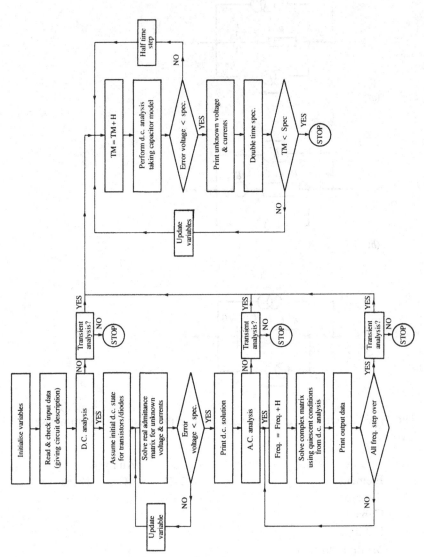

Fig. 8.19. Flow chart of CAIC (after N. D. Arora *et al.*).

topology of the circuit is defined next and the component values are specified using the data statements:

$$R = 100, \ E = 12$$
$$R = 10$$
$$L = 0.15$$
$$C = 0.47E\text{-}6$$

These are written after the appropriate branch as shown below:

B1. $N = (0, 1)$, $R = 100$, $E = 12$: Branch 1 is bounded by node 0 and node 1 and contains a resistor of nominal value 100 ohms and a voltage source of 12 V.

B2. $N = (1, 2)$, $R = 10$: Branch 2 is bounded by node 1 and node 2 and contains a resistor of nominal value 10 ohms.

B3. $N = (2, 3)$, $L = 0.15$: Branch 3 is bounded by node 2 and node 3 and contains an inductor of nominal value 0.15 H.

B4. $N = (3, 0)$, $C = 0.47E\text{-}6$: Branch 4 is bounded by node 3 and node 0 and contains a capacitor with nominal value 0.47 μF.

Frequency = 1000; the frequency of the a.c. source is set to 1000 Hz.

The control statements define the type of analysis that is to be carried out.

First the data and control statements are fed into the computer. The program uses the data to set up the circuit equations and determines which solutions of the equations to perform. The initial design can be altered if the analysis indicates the need for such changes and the new circuit is again analysed. This can be repeated until acceptable performance characteristics are obtained and then the circuit construction can begin.

Figure 8.19 shows the basic flow chart of a program for the *Computer Analysis of Integrated Circuits* (CAIC) developed by Arora *et al.* It can perform nonlinear d.c. analysis, small-signal a.c. analysis and large-signal transient analysis with automatic time-step adjustment. The circuit models for diodes and transistors are predefined and one has to specify only the electrical parameters characterising the device. For transistors a modified version of Ebers–Moll model is used. The model used for junction diodes is shown in Fig. 8.20.

The computer output for the d.c. analysis of an IC operational amplifier CA 3015 is shown in Fig. 8.21 together with measured d.c. voltages at various nodes; see Table 8.1. The computed magnitude response and the phase response of the operational amplifier are shown in Figs. 8.22 and 8.23 respectively. The circuit diagram of an emitter-coupled astable multi-vibrator and its computed transient response are shown in Figs. 8.24 and 8.25 respectively.

Fig. 8.20. Nonlinear model for a junction diode (after N. D. Arora *et al.*).

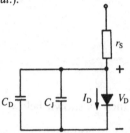

Fig. 8.21. Schematic diagram of CA 3015 operational amplifier.

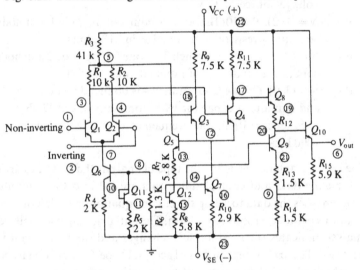

Fig. 8.22. Magnitude response for the circuit of Fig. 8.21 using CAIC (after N. D. Arora *et al.*).

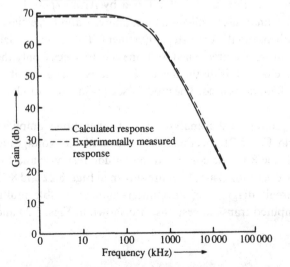

Table 8.1. *Computed and measured d.c. voltages of CA 3015 operational amplifier (after N. D. Arora et al.).*

Node no.	Computed voltage (V)	Measured voltage (V)
1	0.0003	0.0
2	0.0003	0.0
3	0.876	0.8
4	0.875	0.902
5	2.874	2.85
6	0.072	0.082
7	−6.730	−0.674
8	−4.578	−4.238
10	−5.210	−4.931
11	−5.210	−4.932
12	0.202	0.183
13	−0.390	−0.432
14	−2.964	−3.144
15	−3.596	−3.598
16	−3.677	−3.71
17	3.009	3.36
18	3.009	3.160
19	2.375	2.405
20	0.722	0.734
21	−3.598	−3.500
22	6.00	6.00
23	−6.00	−6.00

Fig. 8.23. Phase response for the circuit of Fig. 8.21 using CAIC (after N. D. Arora *et al.*).

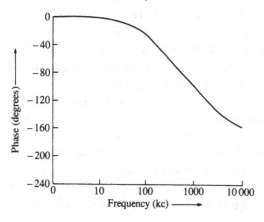

Fig. 8.24. Circuit diagram of an emitter-coupled astable multivibrator.

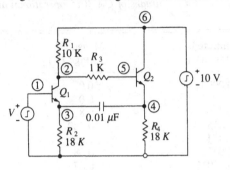

Fig. 8.25. Transient response of the circuit of Fig. 8.24 using CAIC (after N. D. Arora *et al.*).

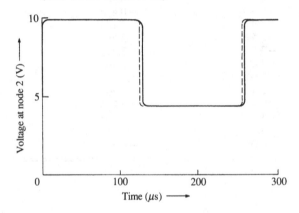

In order to establish the performance characteristics of operational amplifiers, a program known as *Simulation Program with Integrated Circuit Emphasis* (SPICE) can be used (Boyle *et al.*). Table 8.2 compares the performance characteristics of the 8741 type operational amplifier obtained by using this program with that obtained from the Manufacturer's data sheet. These values may be used to determine the element and parameter values of the macromodel of the operational amplifier.

Another computer program IMAG II worked out by Merckel *et al.* at the Applied Mathematics Institute at Grenoble, France, was used to simulate electronic circuits with FETs. Stray capacitances and resistances in the measuring equipment were taken into account. The d.c. transfer curve and also a transient curve for a circuit are shown in Fig. 8.26.

MICRO-CAP II is another well used analogue circuit design and analysis tool. With MICRO-CAP II, one can outline a circuit diagram on a

Table 8.2. *Operational amplifier performance characteristics obtained by using SPICE (after G. R. Boyle, ©197-IEEE).*

	8741 Device-level model	8741 macromodel	LM741 data sheet
C_2 (pF)	30	30	30
S_R^+ (V/μs)	0.9	0.899	0.67
S_R^- (V/μs)	0.72	0.718	0.62
I_B (nA)	256	255	80
I_{Bos} (nA)	0.7	<1	20
V_{os} (mV)	0.299	0.298	1
a_{VD}	$4.17 \cdot 10^5$	$4.16 \cdot 10^5$	$2 \cdot 10^5$
a_{VD} (1 kHz)	$1.219 \cdot 10^3$	$1.217 \cdot 10^3$	10^3
$\Delta\phi$ (°)	16.8	16.3	20
CMRR (dB)	106	106	90
R_{out} (Ω)	566	566	75
R_{o-ac} (Ω)	76.8	76.8	—
I_{SC}^+ (mA)	25.9	26.2	25
I_{SC}^- (mA)	25.9	26.2	25
V^+ (V)	14.2	14.2	14.0
V^- (V)	−12.7	−12.7	−13.5
P_d (mW)	59.4	59.4	—

Fig. 8.26. DC transfer curve and transient curve obtained by using IMAG II (after G. Merckel *et al.*, ©197-IEEE).

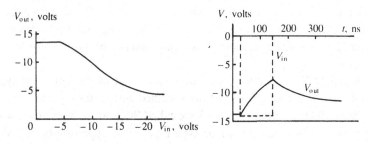

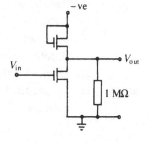

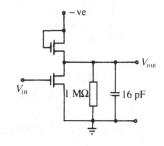

Fig. 8.27. Simplified flow diagram for the MICRO-CAP II.

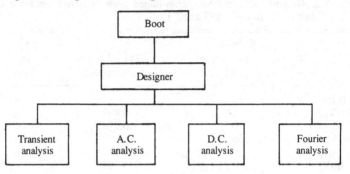

video display monitor. The program automatically creates a netlist suitable for simulation or analysis directly from the graphics image using device models that one defines. MICRO-CAP II then can be used to perform a.c., d.c., transient, or Fourier analysis. The circuit schematic and analysis run may be recorded on a printer in graphic or tabular form.

MICRO-CAP II is supplied with a program diskette and a data diskette. The latter contains several sample circuit files. They are provided as guides to help one learn the system. The data diskette also contains a library file of standard components. In order to use this program one will require an IBM pc, XT or AT computer or equivalent with at least 256k installed random access memory.

Figure 8.27 shows a simplified flow diagram. When the program disk is booted the first program that comes up is the DESIGNER. This is the main program. It allows one to draw circuit diagrams, save and retrieve them from the data diskette, manage the component library and select analysis options.

MICRO-CAP II allows networks to consist of the following types of components. Components 1–6 are specified with a single parametric value such as 1000. Components 7–13 are called Standard Components and are specified by a type number or type label. This type number refers to a library which contains the parameters defining the component.

1. Batteries	2. Resistors
3. Capacitors	4. Inductors
5. User Sources	6. Switches
7. Diodes	8. Bipolar Transistors
9. MOS Devices	10. Operational Amplifiers
11. Sinusoidal Voltage Sources	12. Programmable Voltage Sources
13. Polynomial Sources	

Fig. 8.28. A tuned amplifier circuit and its a.c. analysis obtained by using the MICRO-CAP II.

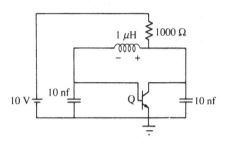

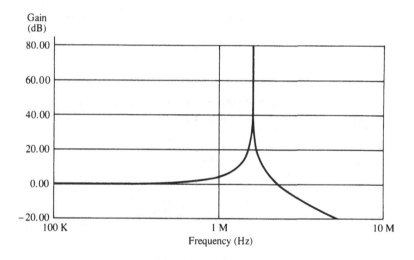

Figure 8.28 shows a tuned amplifier circuit built with an n–p–n type transistor. The a.c. analysis obtained by using the MICRO-CAP II is also shown for a sinusoidal input. The transistor model which has been used for the analysis is shown in Fig. 8.29 and can be compared with the model developed in Section 8.1.1 (see Fig. 8.11).

As another example the frequency response of an infinite gain multiple feedback low-pass filter circuit with the same specification as given in Worked example 3.19 has been obtained with the help of MICRO-CAP II. It is shown in Fig. 8.30 along with the circuit itself.

Fig. 8.29. The bipolar transistor model used for the MICRO-CAP II.

Symbol:

Model:

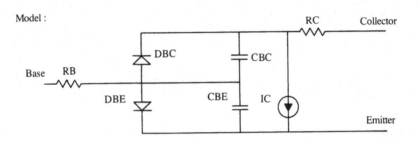

Fig. 8.30. A second-order low-pass filter circuit and its frequency response obtained by using MICRO-CAP II.

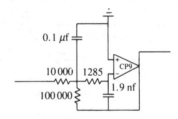

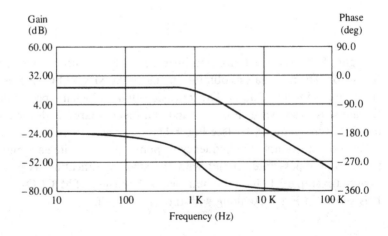

Summary

1. The most widely used model for a transistor is based on the Ebers–Moll model. This replaces the transistors in a circuit enabling one to acquire a set of nonlinear equations in terms of voltages, currents and passive element values. By using correct computation methods for the solution of these equations one can obtain the nonlinear d.c. analysis, small-signal a.c. analysis and large-signal transient analysis of the circuit.

2. The regions of operation of a transistor are defined by the bias voltage on its base-emitter and base-collector junctions. Simplified Ebers–Moll models can be derived for each of these regions and used for machine, or even hand computation.

3. A model for metal-oxide semiconductor type FETs has been derived. These transistors can be more easily modelled in terms of the physical device parameters than those of the bipolar junction transistors. The capacitances in this case may be assumed to be independent of voltages without much loss of accuracy and the bulk series resistances associated with both the source and the drain terminals may be neglected.

4. A model has been described which can accurately simulate the input, intermediate and output stages of an operational amplifier circuit. The parameters of the model may be calculated from the specifications of the operational amplifier supplied by the manufacturers or can be computed from the parameters of discrete components. The design equations for the parameters and element values of the model have been derived. These have been used to the full extent in an example.

5. The circuit analysis techniques are of three categories: d.c. analysis, small-signal a.c. analysis and large-signal transient analysis. The d.c. analysis may be obtained using an iterative numerical method, e.g., Newton–Raphson Method. In this method an initial guess is made of the circuit d.c. values which are then updated. The process continues until the d.c. values on successive iterations converge to within specified tolerance. This algorithm converges rapidly but the disadvantage is that the Jacobean matrix of the set of equations describing the circuit has to be evaluated.

6. The d.c. operating points are used to compute linearised model parameters for the a.c. small signal analysis and initial conditions for the transient analysis.

7. In the process of circuit designs, programs are used to perform repeated d.c., a.c. and transient analysis of a circuit. These include the equation compilers, which generate the necessary equations from the node and element description of the circuit. A few programs and their applications have been mentioned. The selection of a program depends on a particular analysis requirement and computer facilities available to the designer.

Problems

8.1. The transistor in the circuit of Fig. P.8.1 has the following values of some parameters: $\beta_F = 125$, $V_{CE(sat)} = 0.15$ V and $V_{BE(on)} = 0.5$ V. Determine the values of the input voltages at which the cutoff and saturation take place.

Fig. P.8.1

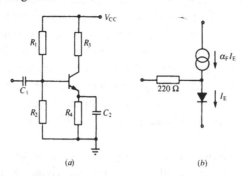

8.2. Plot and dimension the output waveform for the circuit shown in Fig. ex.8.2. The input is a sinusoidal wave of ± 0.25 V magnitude having a frequency of 10 kHz. Given: $V_{BE(on)} = 0.3$ V, $V_{CE(sat)} = 0.1$ V and $\beta_F = 110$.

8.3. The amplifier shown in Fig. P.8.3.(a) is in the forward active mode and the transistor can be replaced by the network shown in Fig. P.8.3(b). Derive a system of equations in matrix form which can be solved to obtain the d.c. conditions with the aid of a computer using the Newton–Raphson technique. The 220 Ω resistor is the base-spreading resistor of the transistor.

Fig. P.8.3

(a)   (b)

8.4. Describe with the aid of diagrams, suitable models for a bipolar junction transistor which can be used to compute non-linear d.c., large signal transient and small signal a.c. values of a circuit comprising such transistors.

Appendix A

Table A1. *Typical h-parameter values for a transistor at* $I_E = 1.3\,mA$

Parameter	Common collector	Common-base	Common-emitter
h_i	$1.1\,k\Omega$	$2.16\,\Omega$	$1.1\,k\Omega$
h_r	0.99	2.9×10^{-4}	2.5×10^{-4}
h_f	-51	-0.98	50
h_o	$25\,\mu A/V$	$0.49\,\mu A/V$	$24\,\mu A/V$

Appendix B

HI506 16-channel CMOS multiplexer.

Table B1. *Electrical characteristics*

$T_A = 25\,^\circ\text{C}$, $V^+ = 15\,\text{V}$, $V^- = -15\,\text{V}$, $\text{GND} = 0$ (Conditions in brackets are for DG 508 only)

Parameter	Symbol	Conditions	HI 506 Typ.	Max.	Units
Drain–Source ON resistance	r_{DSon}[1]	$V_D = 10\,\text{V}$, $V_{in} = 0.8\,\text{V}$ $V_D = -10\,\text{V}$, $V_{in} = 0.8\,\text{V}$	270 270	400 400	Ω Ω
Source OFF leakage current	I_{Soff}	$V_D = \pm 14\,\text{V}$ ($\pm 10\,\text{V}$)	0.03	—	nA
Drain OFF leakage current	I_{Doff}	$V_S = \pm 14\,\text{V}$ ($\pm 10\,\text{V}$)	1.0	—	nA
Channel ON leakage current	I_{Don}[2]	$V_D = V_S \pm 14\,\text{V} \pm 10\,\text{V}$)	1.0	—	nA
Input-transition voltage V_T[3]				0.8/2.4	V
Input (address) current (Input voltage high)	I_{inH} (I_{AH})	$V_{in} = 2.4\,\text{V}$ ($V_A = 2.4\,\text{V}$) $V_{in} = 15\,\text{V}$ ($V_A = 15\,\text{V}$)	— 0.001	— 5	μA μA
Peak input (address) current for transition	I_{in} PEAK (I_{APEAK})	$V_{in} = 1.45\,\text{V}$ ($V_A = 1.45\,\text{V}$) $\pm 10\%$	—	—	μA
Input (address) current (Input voltage low)	I_{inL} (I_L)	$V_{in} = 0\,\text{V}$ ($V_{\text{EN}} = 0\,\text{V}$)	0.001	5	μA
Turn ON time (EN)	t_{on} (EN)		0.3	—	μs
Turn OFF time (EN)	t_{off} (EN)		0.3	—	μs
Source OFF capacitance	C_{Soff}	$V_S = 0$, $V_{in} = 5\,\text{V}$	4	—	pF
Drain OFF capacitance	C_{Doff}	$V_D = 0$, $V_{in} = 5\,\text{V}$	44	—	pF
Channel ON capacitance	$C_{\text{Don}} + C_{\text{Son}}$	$V_D = V_S = V_{\text{IN}} = 0$	—	—	pF
OFF isolation			75	—	dB
Multiplexer switching time	$t_{\text{transition}}$		0.3	—	μs

Table B1. (*cont*).

Parameter	Symbol	Conditions	HI 506 Typ.	Max.	Units
Break before Make time	t_{open}		0.08	—	μs
Operational supply current	$I_{operational}$	$V_{IN} = 0$ ($V_A = 0$, $V_{EN} = 5\,V$)	+3.4 −0.8	+5.0 −2.0	mA
Standby supply current	$I_{standby}$	All channels Off $V_{IN} = 5\,V$ ($V_A = 0$, $V_{EN} = 0\,V$)	+3.4 −0.8	+5.0 −2.0	mA

Notes: [1] r_{DSon} is quoted at $I_D = 1\,mA$ except for DE 508 = 200 μA and DG 303 = 10 mA.
   [2] Leakage from driver into ON switch.
   [3] Where two values are quoted, these are the input Low and High threshold levels.

Table B2. *Absolute maximum ratings*

Parameter	Units	HI 506
Power dissipation (Package)[1]	mW	1200
Supply voltage	V	± 20
V_S or V_D Positive	V	$V^+ + 2\,V$
V_S or V_D Negative	V	$V^- + 2\,V$
V_{in}, V_{ref}, A or EN to Gnd	V	$V^-\,4\,V$ to $V^+ + 4\,V$
Current any terminal except S or D	mA	—
Continuous current S or D	mA	—
Peak current S or D^2	mA	50
Operating temperature	°C	0 to 75
Storage temperature	°C	−65 to +150

Notes: [1] IC soldered into PCB, DG types derate by 6.5 mW/°C above 25 °C. HI 506, derate by 8 mW/°C.
   [2] Pulsed 1 mS, 10% duty cycle maximum.
(RS Components Ltd.)

Appendix C

µA741
FREQUENCY-COMPENSATED OPERATIONAL AMPLIFIER
FAIRCHILD LINEAR INTEGRATED CIRCUITS

GENERAL DESCRIPTION — The µA741 is a high performance monolithic Operational Amplifier constructed using the Fairchild Planar* epitaxial process. It is intended for a wide range of analog applications. High common mode voltage range and absence of latch-up tendencies make the µA741 ideal for use as a voltage follower. The high gain and wide range of operating voltage provides superior performance in integrator, summing amplifier, and general feedback applications.

- NO FREQUENCY COMPENSATION REQUIRED
- SHORT CIRCUIT PROTECTION
- OFFSET VOLTAGE NULL CAPABILITY
- LARGE COMMON MODE AND DIFFERENTIAL VOLTAGE RANGES
- LOW POWER CONSUMPTION
- NO LATCH-UP

ABSOLUTE MAXIMUM RATINGS

Supply Voltage	
µA741A, µA741, µA741E	±22 V
µA741C	±18 V
Internal Power Dissipation (Note 1)	
Metal Can	500 mW
Molded and Hermetic DIP	670 mW
Mini DIP	310 mW
Flatpak	570 mW
Differential Input Voltage	±30 V
Input Voltage (Note 2)	±15 V
Storage Temperature Range	
Metal Can, Hermetic DIP, and Flatpak	−65°C to +150°C
Mini DIP, Molded DIP	−55°C to +125°C
Operating Temperature Range	
Military (µA741A, µA741)	−55°C to +125°C
Commercial (µA741E, µA741C)	0°C to +70°C
Pin Temperature (Soldering)	
Metal Can, Hermetic DIPs, and Flatpak (60 s)	300°C
Molded DIPs (10 s)	260°C
Output Short Circuit Duration (Note 3)	Indefinite

CONNECTION DIAGRAMS

8-PIN METAL CAN
(TOP VIEW)
PACKAGE OUTLINE 5B
PACKAGE CODE H

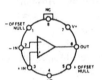

Note: Pin 4 connected to case

ORDER INFORMATION

TYPE	PART NO.
µA741A	µA741AHM
µA741	µA741HM
µA741E	µA741EHC
µA741C	µA741HC

14-PIN DIP
(TOP VIEW)
PACKAGE OUTLINES 6A, 9A
PACKAGE CODES D P

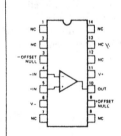

ORDER INFORMATION

TYPE	PART NO.
µA741A	µA741ADM
µA741	µA741DM
µA741E	µA741EDC
µA741C	µA741DC
µA741C	µA741PC

8-PIN MINI DIP
(TOP VIEW)
PACKAGE OUTLINES 6T 9T
PACKAGE CODES R T

ORDER INFORMATION

TYPE	PART NO.
µA741C	µA741TC
µA741C	µA741RC

10-PIN FLATPAK
(TOP VIEW)
PACKAGE OUTLINE 3F
PACKAGE CODE F

ORDER INFORMATION

TYPE	PART NO.
µA741A	µA741AFM
µA741	µA741FM

*Planar is a patented Fairchild process.

FAIRCHILD • µA741

µA741A

ELECTRICAL CHARACTERISTICS: $V_S = \pm 15$ V, $T_A = 25°C$ unless otherwise specified.

CHARACTERISTICS (see definitions)	CONDITIONS		MIN	TYP	MAX	UNITS
Input Offset Voltage	$R_S \leqslant 50\Omega$			0.8	3.0	mV
Average Input Offset Voltage Drift					15	µV/°C
Input Offset Current				3.0	30	nA
Average Input Offset Current Drift					0.5	nA/°C
Input Bias Current				30	80	nA
Power Supply Rejection Ratio	$V_S = +20, -20$; $V_S = -20, +10V$, $R_S = 50\Omega$			15	50	µV/V
Output Short Circuit Current			10	25	40	mA
Power Dissipation	$V_S = \pm 20V$			80	150	mW
Input Impedance	$V_S = \pm 20V$		1.0	6.0		MΩ
Large Signal Voltage Gain	$V_S = \pm 20V$, $R_L = 2k\Omega$, $V_{OUT} = \pm 15V$		50			V/mV
Transient Response	Rise Time			0.25	0.8	µs
(Unity Gain)	Overshoot			6.0	20	%
Bandwidth (Note 4)			.437	1.5		MHz
Slew Rate (Unity Gain)	$V_{IN} = \pm 10V$		0.3	0.7		V/µs
The following specifications apply for $-55°C \leqslant T_A \leqslant +125°C$						
Input Offset Voltage					4.0	mV
Input Offset Current					70	nA
Input Bias Current					210	nA
Common Mode Rejection Ratio	$V_S = \pm 20V$, $V_{IN} = \pm 15V$, $R_S = 50\Omega$		80	95		dB
Adjustment For Input Offset Voltage	$V_S = \pm 20V$		10			mV
Output Short Circuit Current			10		40	mA
Power Dissipation	$V_S = \pm 20V$	$-55°C$			165	mW
		$+125°C$			135	mW
Input Impedance	$V_S = \pm 20V$		0.5			MΩ
Output Voltage Swing	$V_S = \pm 20V$,	$R_L = 10k\Omega$	±16			V
		$R_L = 2k\Omega$	±15			V
Large Signal Voltage Gain	$V_S = \pm 20V$, $R_L = 2k\Omega$, $V_{OUT} = \pm 15V$		32			V/mV
	$V_S = \pm 5V$, $R_L = 2k\Omega$, $V_{OUT} = \pm 2$ V		10			V/mV

NOTES

1. Rating applies to ambient temperatures up to 70°C. Above 70°C ambient derate linearly at 6.3mW/°C for the metal can, 8.3mW/°C for the DIP and 7.1mW/°C for the Flatpak.
2. For supply voltages less than ±15V, the absolute maximum input voltage is equal to the supply voltage.
3. Short circuit may be to ground or either supply. Rating applies to +125°C case temperature or 75°C ambient temperature.
4. Calculate value from: $BW(MHz) = \dfrac{0.35}{Rise\ Time\ (µs)}$

FAIRCHILD • μA741

μA741

ELECTRICAL CHARACTERISTICS: V_S = ±15 V, T_A = 25°C unless otherwise specified.

CHARACTERISTICS (see definitions)	CONDITIONS	MIN	TYP	MAX	UNITS
Input Offset Voltage	R_S < 10 kΩ		1.0	5.0	mV
Input Offset Current			20	200	nA
Input Bias Current			80	500	nA
Input Resistance		0.3	2.0		MΩ
Input Capacitance			1.4		pF
Offset Voltage Adjustment Range			±15		mV
Large Signal Voltage Gain	R_L > 2 kΩ, V_{OUT} = ±10 V	50,000	200,000		
Output Resistance			75		Ω
Output Short Circuit Current			25		mA
Supply Current			1.7	2.8	mA
Power Consumption			50	85	mW
Transient Response (Unity Gain) — Rise time	V_{IN} = 20 mV, R_L = 2 kΩ, C_L < 100 pF		0.3		μs
Overshoot			5.0		%
Slew Rate	R_L > 2 kΩ		0.5		V/μs

The following specifications apply for −55°C < T_A < +125°C:

Input Offset Voltage	R_S < 10 kΩ		1.0	6.0	mV
Input Offset Current	T_A = +125°C		7.0	200	nA
	T_A = −55°C		85	500	nA
Input Bias Current	T_A = +125°C		0.03	0.5	μA
	T_A = −55°C		0.3	1.5	μA
Input Voltage Range		±12	±13		V
Common Mode Rejection Ratio	R_S < 10 kΩ	70	90		dB
Supply Voltage Rejection Ratio	R_S < 10 kΩ		30	150	μV/V
Large Signal Voltage Gain	R_L > 2 kΩ, V_{OUT} = ±10 V	25,000			
Output Voltage Swing	R_L > 10 kΩ	±12	±14		V
	R_L > 2 kΩ	±10	±13		V
Supply Current	T_A = +125°C		1.5	2.5	mA
	T_A = −55°C		2.0	3.3	mA
Power Consumption	T_A = +125°C		45	75	mW
	T_A = −55°C		60	100	mW

TYPICAL PERFORMANCE CURVES FOR μA741A AND μA741

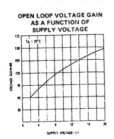

OPEN LOOP VOLTAGE GAIN
AS A FUNCTION OF
SUPPLY VOLTAGE

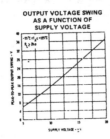

OUTPUT VOLTAGE SWING
AS A FUNCTION OF
SUPPLY VOLTAGE

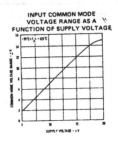

INPUT COMMON MODE
VOLTAGE RANGE AS A
FUNCTION OF SUPPLY VOLTAGE

FAIRCHILD • μA741

μA741E

ELECTRICAL CHARACTERISTICS: V_S = ±15 V, T_A = 25°C unless otherwise specified.

CHARACTERISTICS (see definitions)	CONDITIONS	MIN	TYP	MAX	UNITS
Input Offset Voltage	R_S < 50Ω		0.8	3.0	mV
Average Input Offset Voltage Drift				15	μV/°C
Input Offset Current			3.0	30	nA
Average Input Offset Current Drift				0.5	nA/°C
Input Bias Current			30	80	nA
Power Supply Rejection Ratio	V_S = +10, −20; V_S = +20, −10V, R_S = 50Ω		15	50	μV/V
Output Short Circuit Current		10	25	40	mA
Power Dissipation	V_S = ±20V		80	150	mW
Input Impedance	V_S = ±20V	1.0	6.0		MΩ
Large Signal Voltage Gain	V_S = ±20V, R_L = 2kΩ, V_{OUT} = ±15V	50			V/mV
Transient Response Rise Time			0.25	0.8	μs
(Unity Gain) Overshoot			6.0	20	%
Bandwidth (Note 4)		437	1.5		MHz
Slew Rate (Unity Gain)	V_{IN} = ±10V	0.3	0.7		V/μs
The following specifications apply for 0°C ≤ T_A ≤ 70°C					
Input Offset Voltage				4.0	mV
Input Offset Current				70	nA
Input Bias Current				210	nA
Common Mode Rejection Ratio	V_S = ±20V, V_{IN} = ±15V, R_S = 50Ω	80	95		dB
Adjustment For Input Offset Voltage	V_S = ±20V	10			mV
Output Short Circuit Current		10		40	mA
Power Dissipation	V_S = ±20V			150	mW
Input Impedance	V_S = ±20V	0.5			MΩ
Output Voltage Swing	V_S = ±20V, R_L = 10kΩ	±16			V
	R_L = 2kΩ	±15			V
Large Signal Voltage Gain	V_S = ±20V, R_L = 2kΩ, V_{OUT} = ±15V	32			V/mV
	V_S = ±5V, R_L = 2kΩ, V_{OUT} = ±2 V	10			V/mV

EQUIVALENT CIRCUIT

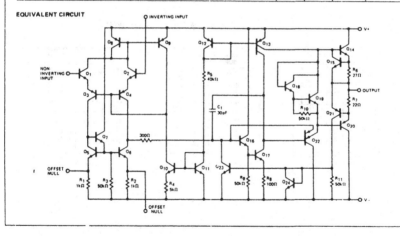

FAIRCHILD • μA741

μA741C
ELECTRICAL CHARACTERISTICS: V_S = ±15 V, T_A = 25°C unless otherwise specified.

CHARACTERISTICS (see definitions)	CONDITIONS		MIN	TYP	MAX	UNITS
Input Offset Voltage	R_S < 10 kΩ			2.0	6.0	mV
Input Offset Current				20	200	nA
Input Bias Current				80	500	nA
Input Resistance			0.3	2.0		MΩ
Input Capacitance				1.4		pF
Offset Voltage Adjustment Range				±15		mV
Input Voltage Range			±12	±13		V
Common Mode Rejection Ratio	R_S < 10 kΩ		70	90		dB
Supply Voltage Rejection Ratio	R_S < 10 kΩ			30	150	μV/V
Large Signal Voltage Gain	R_L ≥ 2 kΩ, V_{OUT} = ±10 V		20,000	200,000		
Output Voltage Swing	R_L ≥ 10 kΩ		±12	±14		V
	R_L ≥ 2 kΩ		±10	±13		V
Output Resistance				75		Ω
Output Short Circuit Current				25		mA
Supply Current				1.7	2.8	mA
Power Consumption				50	85	mW
Transient Response (Unity Gain)	Rise time	V_{IN} = 20 mV, R_L = 2 kΩ, C_L < 100 pF		0.3		μs
	Overshoot			5.0		%
Slew Rate	R_L ≥ 2 kΩ			0.5		V/μs

The following specifications apply for 0°C < T_A < +70°C:

Input Offset Voltage					7.5	mV
Input Offset Current					300	nA
Input Bias Current					800	nA
Large Signal Voltage Gain	R_L ≥ 2 kΩ, V_{OUT} = ±10 V		15,000			
Output Voltage Swing	R_L ≥ 2 kΩ		±10	±13		V

TYPICAL PERFORMANCE CURVES FOR μA741E AND μA741C

OPEN LOOP VOLTAGE GAIN
AS A FUNCTION OF
SUPPLY VOLTAGE

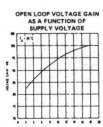

OUTPUT VOLTAGE SWING
AS A FUNCTION OF
SUPPLY VOLTAGE

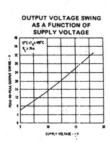

INPUT COMMON MODE
VOLTAGE RANGE AS A
FUNCTION OF SUPPLY VOLTAGE

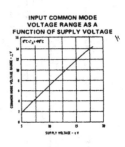

FAIRCHILD • μA741

TYPICAL PERFORMANCE CURVES FOR μA741A, μA741, μA741E AND μA741C

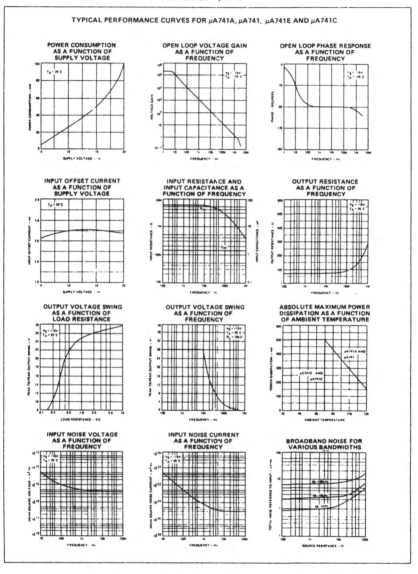

POWER CONSUMPTION AS A FUNCTION OF SUPPLY VOLTAGE

OPEN LOOP VOLTAGE GAIN AS A FUNCTION OF FREQUENCY

OPEN LOOP PHASE RESPONSE AS A FUNCTION OF FREQUENCY

INPUT OFFSET CURRENT AS A FUNCTION OF SUPPLY VOLTAGE

INPUT RESISTANCE AND INPUT CAPACITANCE AS A FUNCTION OF FREQUENCY

OUTPUT RESISTANCE AS A FUNCTION OF FREQUENCY

OUTPUT VOLTAGE SWING AS A FUNCTION OF LOAD RESISTANCE

OUTPUT VOLTAGE SWING AS A FUNCTION OF FREQUENCY

ABSOLUTE MAXIMUM POWER DISSIPATION AS A FUNCTION OF AMBIENT TEMPERATURE

INPUT NOISE VOLTAGE AS A FUNCTION OF FREQUENCY

INPUT NOISE CURRENT AS A FUNCTION OF FREQUENCY

BROADBAND NOISE FOR VARIOUS BANDWIDTHS

FAIRCHILD • μA741

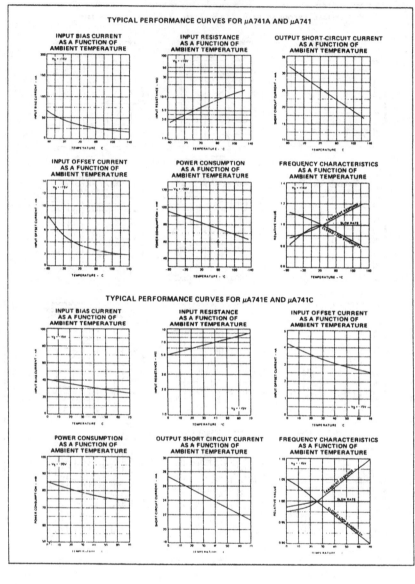

FAIRCHILD • μA741

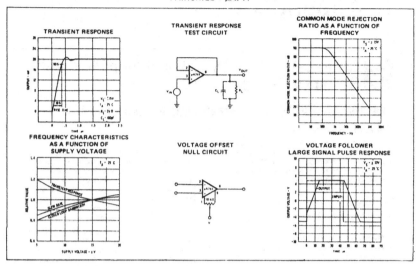

(National Semiconductors Ltd.)

Appendix D

LM118/LM218/LM318 Operational Amplifiers

General Description

The LM118 series are precision high speed operational amplifiers designed for applications requiring wide bandwidth and high slew rate. They feature a factor of ten increase in speed over general purpose devices without sacrificing DC performance.

Features

- 15 MHz small signal bandwidth
- Guaranteed 50V/μs slew rate
- Maximum bias current of 250 nA
- Operates from supplies of ±5V to ±20V
- Internal frequency compensation
- Input and output overload protected
- Pin compatible with general purpose op amps

The LM118 series has internal unity gain frequency compensation. This considerably simplifies its application since no external components are necessary for operation. However, unlike most internally

compensated amplifiers, external frequency compensation may be added for optimum performance For inverting applications, feedforward compensation will boost the slew rate to over 150V/μs and almost double the bandwidth. Overcompensation can be used with the amplifier for greater stability when maximum bandwidth is not needed. Further, a single capacitor can be added to reduce the 0.1% settling time to under 1 μs.

The high speed and fast settling time of these op amps make them useful in A/D converters, oscillators, active filters, sample and hold circuits, or general purpose amplifiers. These devices are easy to apply and offer an order of magnitude better AC performance than industry standards such as the LM709.

The LM218 is identical to the LM118 except that the LM218 has its performance specified over a −25°C to +85°C temperature range. The LM318 is specified from 0°C to +70°C.

Schematic and Connection Diagrams

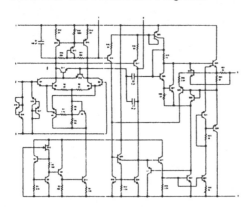

Dual-In-Line Package

Order Number LM118J, LM218J
or LM318J
See NS Package J14A

Metal Can Package*

*Pin connections shown on schematic diagram
and typical applications are for TO-5 package.

Order Number LM118H, LM218H
or LM318H
See NS Package H08C

Dual-In-Line Package

Order Number LM118J-8,
LM218J-8 or LM318J-8
See NS Package J08A
Order Number LM318N
See NS Package N08B

Absolute Maximum Ratings

Supply Voltage	±20V
Power Dissipation (Note 1)	500 mW
Differential Input Current (Note 2)	±10 mA
Input Voltage (Note 3)	±15V
Output Short-Circuit Duration	Indefinite
Operating Temperature Range	
LM118	$-55°C$ to $+125°C$
LM218	$-25°C$ to $+85°C$
LM318	$0°C$ to $+70°C$
Storage Temperature Range	$-65°C$ to $+150°C$
Lead Temperature (Soldering, 10 seconds)	$300°C$

Electrical Characteristics (Note 4)

PARAMETER	CONDITIONS	LM118/LM218			LM318			UNITS
		MIN	TYP	MAX	MIN	TYP	MAX	
Input Offset Voltage	$T_A = 25°C$		2	4		4	10	mV
Input Offset Current	$T_A = 25°C$		6	50		30	200	nA
Input Bias Current	$T_A = 25°C$		120	250		150	500	nA
Input Resistance	$T_A = 25°C$	1	3		0.5	3		MΩ
Supply Current	$T_A = 25°C$		5	8		5	10	mA
Large Signal Voltage Gain	$T_A = 25°C$, $V_S = ±15V$ $V_{OUT} = ±10V$, $R_L \geq 2\,k\Omega$	50	200		25	200		V/mV
Slew Rate	$T_A = 25°C$, $V_S = ±15V$, $A_V = 1$	50	70		50	70		V/μs
Small Signal Bandwidth	$T_A = 25°C$, $V_S = ±15V$		15			15		MHz
Input Offset Voltage				6			15	mV
Input Offset Current				100			300	nA
Input Bias Current				500			750	nA
Supply Current	$T_A = 125°C$		4.5	7				mA
Large Signal Voltage Gain	$V_S = ±15V$, $V_{OUT} = ±10V$ $R_L \geq 2\,k\Omega$	25			20			V/mV
Output Voltage Swing	$V_S = ±15V$, $R_L = 2\,k\Omega$	±12	±13		±12	±13		V
Input Voltage Range	$V_S = ±15V$	±11.5			±11.5			V
Common-Mode Rejection Ratio		80	100		70	100		dB
Supply Voltage Rejection Ratio		70	80		65	80		dB

Note 1: The maximum junction temperature of the LM118 is 150°C, the LM218 is 110°C, and the LM318 is 110°C. For operating at elevated temperatures, devices in the TO-5 package must be derated based on a thermal resistance of 150°C/W, junction to ambient, or 45°C/W, junction to case. The thermal resistance of the dual-in-line package is 100°C/W, junction to ambient.

Note 2: The inputs are shunted with back-to-back diodes for overvoltage protection. Therefore, excessive current will flow if a differential input voltage in excess of 1V is applied between the inputs unless some limiting resistance is used.

Note 3: For supply voltages less than ±15V, the absolute maximum input voltage is equal to the supply voltage.

Note 4: These specifications apply for ±5V $\leq V_S \leq$ ±20V and $-55°C \leq T_A \leq$ +125°C,(LM118), $-25°C \leq T_A \leq$ +85°C (LM218), and $0°C \leq T_A \leq$ +70°C (LM318). Also, power supplies must be bypassed with 0.1μF disc capacitors.

Typical Performance Characteristics LM118, LM218

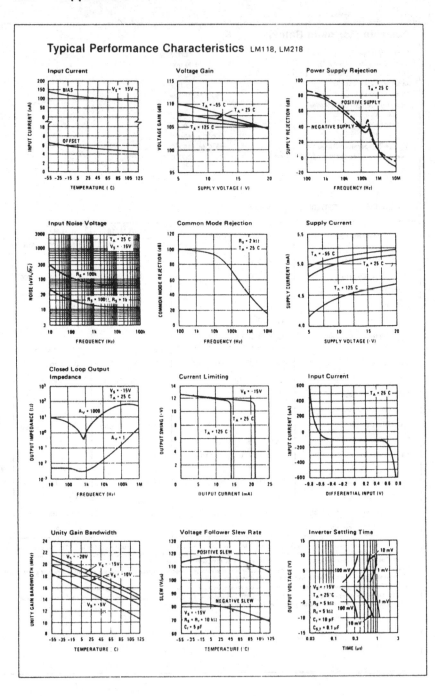

Typical Performance Characteristics LM118, LM218 (Continued)

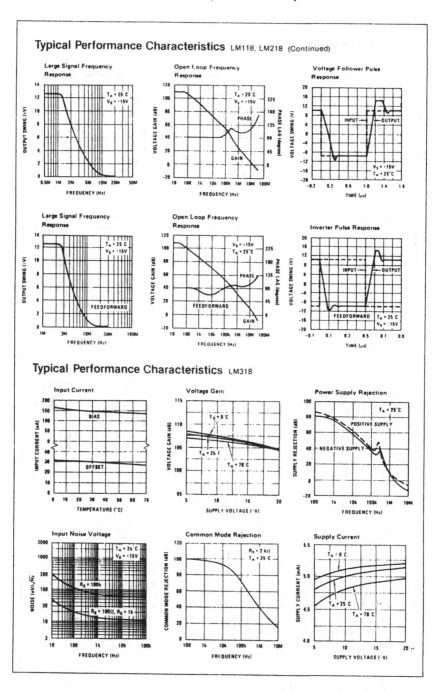

Typical Performance Characteristics LM318 (Continued)

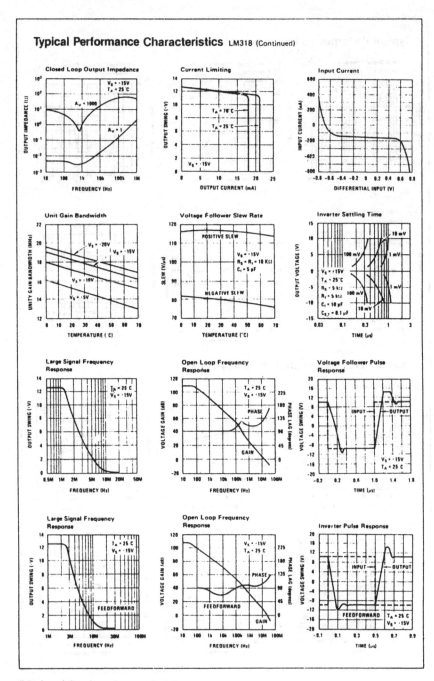

(National Semiconductors Ltd.)

Appendix E

Polynomials for low-pass filters.

Butterworth Low - Pass Filters

n	Polynomial
1	$s+1$
2	$s^2 + \sqrt{2}\,s + 1$
3	$s^3 + 2s^2 + 2s + 1 = (s+1)(s^2 + s + 1)$
4	$s^4 + 2.613s^3 + 3.414s^2 + 2.613s + 1 = (s^2 + 0.765s + 1)(s^2 + 1.848s + 1)$

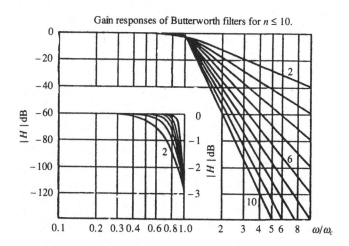

Gain responses of Butterworth filters for $n \le 10$.

Bessel (or Thompson) Low-Pass Filters

n	Polynomial
1	$s+1$
2	s^2+3s+3
3	$s^3+6s^2+15s+15=(s+2.322)(s^2+3.678s+6.460)$
4	$s^4+10s^3+45s^2+105s+105=(s^2+5.792s+9.140)(s^2+4.208s+11.488)$

Gain responses for Thompson (Bessel) filters for $n \leq 10$.

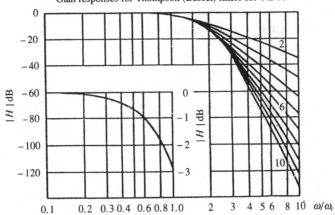

Chebychev Low - Pass Filters

n	Polynomial

1/2-dB ripple ($\varepsilon = 0.3493$)

1   $s + 2.863$

2   $s^2 + 1.425s + 1.516$

3   $s^3 + 1.253s^2 + 1.535s + 0.716 = (s + 0.626)(s^2 + 0.626s + 1.142)$

4   $s^4 + 1.197s^3 + 1.717s^2 + 1.025s + 0.379 = (s^2 + 0.351s + 1.064)(s^2 + 0.845s + 0.356)$

1-dB ripple ($\varepsilon = 0.5088$)

1   $s + 1.965$

2   $s^2 + 1.098s + 1.103$

3   $s^3 + 0.988s^2 + 1.238s + 0.491 = (s + 0.494)(s^2 + 0.490s + 0.994)$

4   $s^4 + 0.953s^3 + 1.454a^2 + 0.743s + 0.276 = (s^2 + 0.279s + 0.987)(s^2 + 0.674s + 0.279)$

2-dB ripple ($\varepsilon = 0.7648$)

1   $s + 1.308$

2   $s^2 + 0.804s + 0.637$

3   $s^3 + 0.738s^2 + 1.022s + 0.327 = (s + 0.402)(s^2 + 0.396s + 0.886)$

4   $s^4 + 0.716s^3 + 1.256s^2 + 0.517s + 0.206 = (s^2 + 0.210s + 0.928)(s^2 + 0.506s + 0.221)$

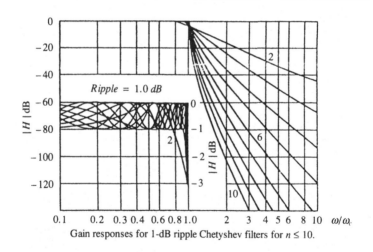

Gain responses for 1-dB ripple Chetyshev filters for $n \leq 10$.

Appendix F

Standard design table for low pass filters.

Table F1.

L and C values for normalised low pass filters. $R = 1\,\Omega$, $\omega_{co} = 1\,\text{rad.s}^{-1}$

	Butterworth	Chebyshev 0.1 dB ripple	Bessel
$\mathscr{C}$	1.4142 F	1.3911 F	1.3617 F
$\mathscr{L}$	0.7071 H	0.8191 H	0.4539 H

$$L = \frac{\mathscr{L}R}{\omega_{co}} \qquad C = \frac{\mathscr{C}}{R\omega_{co}}$$

Appendix G

MF10-type switched capacitor filters

Connection diagram

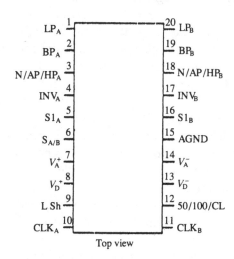

Top view

Pin description

LP, BP, N/AP/HP	These are the lowpass, bandpass, notch or allpass or highpass outputs of each 2nd order section. The LP and BP outputs can sink typically 1 mA and source 3 mA. The N/AP/HP output can typically sink and source 1.5 mA and 3 mA, respectively.
INV	This is the inverting input of the summing op amp of each filter. The pin has static discharge protection.
S1	S1 is a signal input pin used in the allpass filter configurations (see modes of operation 4 and 5). The pin should be driven with a source impedance of less than 1 kΩ.

$S_{A/B}$ It activates a switch connecting one of the inputs of the filter's 2nd summer either to analog ground ($S_{A/B}$ low to V_A^-) or to the lowpass output of the circuit ($S_{A/B}$ high to V_A^+). This allows flexibility in the various modes of operation of the IC. $S_{A/B}$ is protected against static discharge.

V_A^+, V_D^+ Analog positive supply and digital positive supply. These pins are internally connected through the IC substrate and therefore V_A^+ and V_D^+ should be derived from the same power supply source. They have been brought out separately so they can be bypassed by separate capacitors, if desired. They can be externally tied together and bypassed by a single capacitor.

V_A^-, V_D^- Analog and digital negative supply respectively. The same comments as for V_A^+ and V_D^+ apply here.

L Sh   Level shift pin; it accommodates various clock levels with dual or single supply operation. With dual ± 5 V supplies, the MF10 can be driven with CMOS clock levels (± 5 V) and the L Sh pin should be tied either to the system ground or to the negative supply pin. If the same supplies as above are used but TTL clock levels, derived from 0 V to 5 V supply, are only available, the L Sh pin should be tied to the system ground. For single supply operation (0 V and 10 V) the V_D^-, V_A^- pins should be connected to the system ground, the AGND pin should be biased at 5 V and the L Sh pin should also be tied to the system ground. This will accommodate both CMOS and TTL clock levels.

CLK (A or B)   Clock inputs for each switched capacitor filter building block. They should both be of the same level (TTL or CMOS). The level shift (L Sh) pin description discusses how to accommodate their levels. The duty cycle of the clock should preferably be close to 50% especially when clock frequencies above 200 kHz are used. This allows the maximum time for the op amps to settle which yields optimum filter operation.

50/100/CL   By tying the pin high a 50:1 clock to filter centre frequency operation is obtained. Tying the pin at mid supplies (i.e., analog ground with dual supplies) allows the filter to operate at a 100:1 clock to centre frequency ratio. When the pin is tied low, a simple current limiting circuitry is triggered to limit the overall supply of current down to about 2.5 mA. The filtering action is then aborted.

AGND   Analog ground pin; it should be connected to the system ground for dual supply operation or biased at mid supply for single supply operation. The positive inputs of the filter op amps are connected to the AGND pin so 'clean' ground is mandatory. The AGND pin is protected against static discharge.

Definition of terms

f_{CLK}: the switched capacitor filter external clock frequency.

f_0: centre of frequency of the second order function complex pole pair. f_0 is measured at the bandpass output of each 1/2 MF10, and it is the frequency of the bandpass peak occurrence (Fig. G-1).

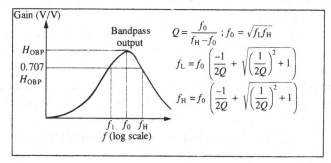

Q: quality factor of the 2nd order function complex pole pair. Q is also measured at the bandpass output of each 1/2 MF10 and it is the ratio of f_0 over the -3 dB bandwidth of the 2nd order bandpass filter, Fig. G-1. The value of Q is not measured at the lowpass or highpass outputs of the filter, but its value relates to the possible amplitude peaking at the above outputs.

H_{OBP}: the gain in (V/V) of the bandpass output at $f=f_0$.

H_{OLP}: the gain in (V/V) of the lowpass output of each 1/2 MF10 at $f\rightarrow 0$ Hz, Fig. G-2.

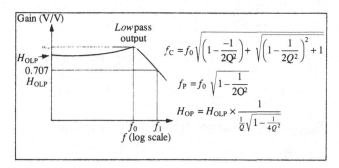

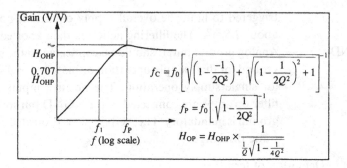

H_{OHP}: the gain in (V/V) of the highpass output of each 1/2 MF10 as $f \rightarrow f_{CLK}/2$, Fig. G-3.

Q_z: the quality factor of the 2nd order function complex zero pair, if any. (Q_z is a parameter used when an allpass output is sought and unlike Q it cannot be directly measured.)

f_z: the centre frequency of the 2nd order function complex zero pair, if any. If f_z is different from f_0, and if the Q_z is quite high it can be observed as a notch frequency at the allpass output.

f_{notch}: the notch frequency observed at the notch output(s) of the MF10.

H_{ON_1}: the notch output gain as $f \rightarrow 0\,\text{Hz}$.

H_{ON_2}: the notch output gain as $f \rightarrow f_{CLK}/2$.

Mode 1a

A configuration that gives good output dynamics together with the provision for high Q in bandpass designs. The notch output features equal gain above and below the notch frequency.

Design equations
Outputs—Notch
     Bandpass
     Lowpass
f_0 = centre frequency of the complex pole pair

$$= \frac{f_{CLK}}{100} \quad \text{or} \quad \frac{f_{CLK}}{50}$$

f_{notch} = centre frequency of the imaginary zero pair

$$= f_0$$

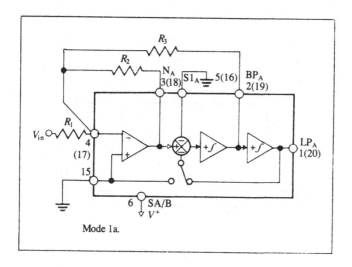

Mode 1a.

$$H_{OLP} = \text{Lowpass gain (as } f \to 0) = -\frac{R_2}{R_1}$$

$$H_{OBP} = \text{Bandpass gain (as } f = f_0) = -\frac{R_3}{R_1}$$

$$H_{ON_1} = H_{ON_2} = -\frac{R_2}{R_1}$$

$$Q = f_0/\text{BW} = \frac{R_3}{R_2}$$

Circuit dynamics
$$H_{OBP} = H_{OLP} \times Q = H_{ON} \times Q$$

$H_{OLP(Peak)} = Q \times H_{OLP}$ (If the DC gain of LP output is too high, a high Q value could cause clipping at the lowpass output resulting in gain non-linearity and distortion at the bandpass output.)

Mode 3a

The most versatile mode of operation, this configuration is the classical state variable filter implemented with only 4 external resistors.

Clock to centre frequency ratio can be externally tuned either above or below the 100:1 or 50:1 values and thus make it suitable for multiple stage Chebyshev filters controlled by a single clock.

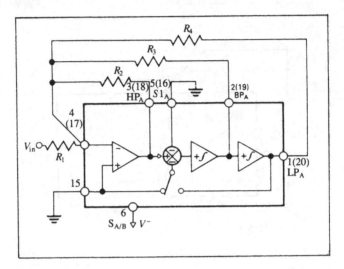

Design Equations

Outputs—Highpass
　　　　　Bandpass
　　　　　Lowpass

$$f_0 = \frac{f_{CLK}}{100} \times \sqrt{\frac{R_2}{R_4}} \text{ or } \frac{f_{CLK}}{50} \times \sqrt{\frac{R_2}{R_4}}$$

Q = quality factor of the complex pole pair

$$= \sqrt{\frac{R_2}{R_4}} \times \frac{R_3}{R_2}$$

$$H_{OHP} = \text{Highpass gain as} \left(f \to \frac{f_{CLK}}{2}\right) = -\frac{R_2}{R_1}$$

$$H_{OBP} = \text{Bandpass gain (at } f=f_0) = -\frac{R_3}{R_1}$$

$$H_{OLP} = \text{Lowpass gain (as } f \to 0) = -\frac{R_4}{R_1}$$

Circuit dynamics

$$\frac{R_2}{R_4} = \frac{H_{OHP}}{H_{OLP}}$$

$$H_{OBP} = Q \times \sqrt{H_{OHP} \times H_{OLP}}$$

$$H_{OLP(peak)} = Q \times H_{OLP} \text{ (for high } Q\text{s)}$$

$$H_{OHP(peak)} = Q \times H_{OHP} \text{ (for high } Q\text{s)}$$

(RS Components Ltd)

Glossary of symbols

A_{OL} = open-loop gain of operational amplifier
$A_{d(s)}$ = single-ended differential voltage gain
$A_{d(d)}$ = double-ended differential voltage gain
A_I = current gain of an amplifier
A_V = voltage gain of an amplifier
α_F = common-base forward current gain
α_R = common-base reverse current gain

β = feedback fraction, feedback factor, common-emitter current gain
BW = bandwidth

$C_{S(OFF)}$ = channel input capacitance
$C_{D(OFF)}$ = channel output capacitance
$C_{DS(OFF)}$ = capacitance between 'OFF' switch terminals
$c(t)$ = sinusoidal carrier wave
C_{DC}, C_{DE} = diffusion capacitances
C_{JX} = junction capacitance
C_{sub} = substrate capacitance
C_{ox} = thin-oxide capacitance per unit area
CMMR = common-mode rejection ratio

ε_n = individual error source
E_g = energy gap of semiconductor material
$\Delta E_{os}/\Delta T$ = offset voltage drift with temperature
$\Delta E_{os}/\Delta V_S$ = offset voltage drift with power supply variation
$\Delta E_{os}/\Delta t$ = offset voltage drift with time

f_c = clock frequency
f_L = lower cutoff frequency
f_U = upper cutoff frequency
f_0 = centre frequency

$\Delta f_{3\,dB} = 3$ dB bandwidth

FSR = full-scale reading

ϕ = barrier potential

g_m = transconductance

g_{mF} = forward transconductance

g_{mR} = reverse transconductance

G = conductance

h_i, h_r, h_f, h_o = hybrid parameters of a two-port device

i_c = total instantaneous collector current

I_c = quiescent collector current

i_{CQ} = instantaneous variation of collector current from the quiescent value

i_B = total instantaneous base current

I_B = quiescent base current

i_{BQ} = instantaneous variation of base current from the quiescent value

I_{OFF} = diode leakage current

I_D = diode current

I_S = diode reverse saturation current

I_o = output current

$\bar{I}_o$ = complementary output current

$I_{leakage}$, I_{DOFF} = leakage current through open switches

I_{DON} = channel 'ON' leakage current flowing from driver into 'ON' switch

I_{ES} = emitter-base saturation current

I_{CS} = collector-base saturation current

I_E = emitter current

I_C = collector current

I_{EO} = reverse biased emitter current with collector terminal open

I_{CO} = reverse biased collector current with emitter terminal open

I_{os} = input offset current

I_b = bias current at inverting/noninverting terminal

$\Delta I_{os}/\Delta T$ = offset current drift with temperature

I_{DS} = drain-source current

I_{DSS} = drain-source saturation current

$I_{o(max)}$ = maximum output current

$I_{D(max)}$ = maximum diode current

k = Boltzmann's constant

K_m = multiplier constant

K_{VCO} = gain of conversion of VCO

LSB = least significant bit

m = gradient factor of p–n junction
m_f = index of frequency modulation
MSB = most significant bit
μ_0 = carrier mobility

ω_{LPF} = cutoff frequency of low-pass filter
ω_L = lock range
ω_{AQ} = acquisition range
ω_c = carrier frequency

P_d = power dissipation

q = electron charge
Q = quality factor
Q_{eff} = effective Q
Q_{DC}, Q_{DE} = mobile charges

r_{in} = effective input resistance of a Darlington pair
r_e = effective output resistance of a Darlington pair
r_π = base-emitter resistance
r_x = base spreading resistance
r_o = output resistance
r_{on} = resistance of switches when 'ON'

s = a complex variable having the form $\alpha + j\omega$
S = slew rate

t_A = access time
t_e = settling time
t_d = intentional delay time
T = absolute temperature in K
T_a = ambient temperature
T_j = junction temperature
θ_c = conduction angle
θ_{cs} = thermal resistance between case and sink
θ_{sa} = thermal resistance between sink and ambient
θ_{jc} = thermal resistance between junction and case
τ_F, τ_R = forward and reverse transit time of charges

V_A = Early voltage
v_C = total instantaneous collector voltage
V_C = quiescent collector voltage
v_{CQ} = instantaneous variation of collector voltage from the quiescent value
V_D = diode voltage
V_{BB} = d.c. biasing voltage
$V_{BE(ON)}$ = base-emitter 'ON' voltage
V_{BE} = base-emitter voltage
V_{BC} = base-collector voltage
V_{CC} = positive d.c. power supply
V_{EE} = negative d.c. power supply
V_H = positive saturation voltage of comparator
V_L = negative saturation voltage of comparator
V_x = analogue voltage
V_R = analogue reference voltage
V_{io} = input offset voltage
V_G = effective gate–source voltage
V_{DS} = drain–source voltage
V_{DSS} = drain–source saturation voltage

W = base width

X_d = binary code

Y_o = output admittance

Z = channel width
Z_i = input impedance

Answers to problems

Chapter 2

2.1. $1.035\,\text{mA}$; terminal C_1; $36.36\,\text{mV}$.

2.2. -160; $319.8\,\text{k}\Omega$; $0.997\,\text{mA}$.

2.3. $1.682\,\text{mA}$.

2.4. $0.5\,\text{mA}$; $10\,\text{V}$.

2.5. 54.

2.6. $2.8\,\text{mA}$; $2.8\,\text{mA}$; $2.8\,\text{mA}$; $18.6\,\text{V}$; $6.2\,\text{V}$; $12.08\,\text{V}$.

2.8. 0.806 or 1.024.

2.9. $1.25\,\text{mA}$.

2.10. The answer is not unique.

2.11. $5.02\,\text{V}$.

2.12. $765.1\,\Omega$.

2.13. $1\,\text{MHz}$; 100.52.

2.14. The transistor should be able to dissipate $12\,\text{W}$; $6\,\text{W}$.

2.15. -0.02; $30\,\text{V}$; $10\,\text{V}$ peak; $1\,\text{W}$; 25%.

2.16. $11.78\,\text{V}$; $3.4\,\text{A}$.

2.17. $0.1\,\text{W}$; 20.8%.

2.18. $4.94\,\text{V}$; $422\,\text{pF}$.

2.19. $16\,\text{W}$.

2.20. $1\,°\text{C/W}$.

2.21. $27.5\,°\text{C}$.

2.22. $150\,°\text{C}$.

Chapter 3

3.1. $2.5\,\mu s$; $0.8\,V$.

3.2. $0.5997\,V/\mu s$.

3.3. $49.736\,kHz$.

3.4. $\pm 15.12\,mV$.

3.5. $5\,mV$; $7.47\,mV$; $5.71\,mV$; $979\,\Omega$.

3.6. $\pm 19.64\,nA$; $\pm 100.89\,nA$.

3.7. $6.46\,\mu V/^{\circ}C$.

3.8. $89.28\,\Omega$; $10.25\,V$.

3.9. $423.6\,hrs$.

3.10. $\pm 286\,mV$.

3.11. $2.2\,M\Omega$.

3.12. $0.143\,\mu F$; $111\,\Omega$.

3.13. $0.35\,\mu F$; $150\,\Omega$.

3.14. $5.41\,V$; $1.564\,V$.

3.18. $150\,\mu F$; $2.2\,k\Omega$; $2.2\,k\Omega$; $2.2\,k\Omega$; $1.1\,k\Omega$; $26.4\,k\Omega$.

3.19. $55.6\,\mu F$; $1\,k\Omega$; $1\,k\Omega$; $1\,k\Omega$; $500\,\Omega$; $10\,k\Omega$.

3.20. $2.169\,mV$.

3.21. $-0.348\,V$; $-0.464\,V$.

3.22. $8.17\,k\Omega$.

3.25. 12.5; $472\,Hz$; $2.86\,nF$.

3.26. $1.9\,kHz$; $31.8\,pF$; $76.9\,pF$.

3.27. $5\,k\Omega$; $0.1\,\mu F$; $1.5\,k\Omega$; $50\,k\Omega$; $3.37\,nF$.

3.28. $113\,nF$; $2.67\,nF$; $854.7\,k\Omega$.

3.29. $0.01\,\mu F$; $2.67\,k\Omega$; $0.01\,\mu F$; $1\,nF$; $59.1\,k\Omega$; $0.98\,V_{P-P}$.

3.30. 6.14; 8.48.

3.31. $6.89\,k\Omega$; $163\,\Omega$; $1\,nF$; $1\,nF$; $15.9\,k\Omega$.

3.32. First stage: $1\,k\Omega$; $10\,k\Omega$; $7.65\,k\Omega$; second stage: $1\,k\Omega$; $10\,k\Omega$; $18.48\,k\Omega$.

3.33. 2.747 kHz.

3.34. $C = 0.01\,\mu\text{F}$; $R_1 = 1\,\text{k}\Omega$; $R_2 = 10\,\text{k}\Omega$; $R_s = 25.6\,\text{k}\Omega$; $R_M = 11\,\text{k}\Omega$.

Chapter 4

4.1. $4.95 \times 10^4 / (-\omega^2 + \text{j}12.56 \times 10^3\omega + 39.4 \times 10^6)$.

4.2. 636.9 Hz.

4.3. 1.33 pF to 3.78 nF.

4.4. 925 μH.

4.5. 41.4 μH to 117 mH.

4.7. $C = 100\,\text{nF}$; $R = 433.2\,\Omega$; $R' = 452.8\,\Omega$.

4.8. $C = 0.01\,\mu\text{F}$; $R = 776\,\Omega$; $R_1 = 10\,\text{k}\Omega$; $R_2 = 20\,\text{k}\Omega$.

4.9. $C = 6.5\,\text{pF}$ to 65 pF; $R = 2.58\,\text{M}\Omega$ to 244.7 kΩ; $R_1 = 10\,\text{k}\Omega$; $R_2 = 20\,\text{k}\Omega$.

4.10. 464.3 to 465.9 kHz.

Chapter 5

5.1. $6.37 \times 10^4\,\text{rad.s}^{-1}$.

5.2. $2.16 \times 10^3\,\text{rad.s}^{-1}$.

5.3. $13.662 \times 10^3\,\text{rad.s}^{-1}$.

5.4. 1.65 volts.

5.5. 2.34 volts.

5.6. $-19.3°$.

5.7. $-13.6°$.

5.9. 50 000 rad.s^{-1}; 11 000 rad.s^{-1}.

Chapter 6

6.1. 800 to 809.85 kHz; 810.15 to 820 kHz.

6.2. 287.5 to 288.8 kHz; passband: 287.5 to 292.5 kHz.

6.3. $14.14\cos 1445 \times 10^3 t + 3.535\cos 37.7 \times 10^3 t \cos 1445 \times 10^3 t$.

6.4. 9%.

6.5. (a) 37 kHz, (b) 74 kHz.

6.8. 7.14; 57 kHz.

6.9. $\omega = 50 \times 10^3 + 20 \times 10^3 \cos 4\pi \times 10^3 t$.

6.10. $3.375 \, \mu s$.

Chapter 7

7.1. 10 volts, 5.94 volts.

7.2. 98.28%.

7.3. $45.18 \, k\Omega$.

7.4. 2.9882 mA, 1.9921 mA.

7.5. 3.0159 MHz.

7.6. $2441.4 \, V.s^{-1}$.

7.7. MSB = 4.5 volts, LSB = 0.140625 volts, 110001, 0.136%.

7.8. Successive approximation type.

7.9. 18.07 ms.

7.11. 0.053 mV.

7.12. 1026000 samples per second.

7.13. 0.036 mV, $0.22 \, \mu s$.

7.14. 10763 samples per second.

7.15. yes.

7.16. $113.06 \, \mu s$; $3.64 \, mV.s^{-1}$.

7.17. $2.82 \, \mu s$; $851 \, mV.s^{-1}$; 1.29 ms.

7.18. 11.11 nF; 1.36 nF.

Chapter 8

8.1. 0.5 V; 1.027 V.

References

Books
General

1.   H. Ahmed & P.J. Spreadbury, *Electronics for Engineers: An Introduction*, Cambridge University Press, Cambridge, 1973.
2.   T. F. Bogart, *Electronic Devices and Circuits*, Merrill Publishing Company, Ohio, 1986.
3.   P. M. Chirlian, *Analysis and Design of Integrated Electronic Circuits*, Harper & Row Publishers, NY, 1981.
4.   D. J. Comer, *Modern Electronic Circuit Design*, Addison-Wesley Publishing Company Inc., Phillippines, 1977.
5.   D. J. Comer, *Electronic Design with Integrated Circuits*, Addison-Wesley Publishing Company Inc., Phillippines, 1981.
6.   G. M. Glasford, *Analog Electronic Circuits*, Prentice-Hall International Inc., NJ, 1986.
7.   P. R. Gray & R. C. Meyer, *Analog Integrated Circuits*, Wiley, New York, 1984.
8.   B. Grob, *Electronic Circuits and Applications*, McGraw-Hill International Book Company, 1982.
9.   P. Horowitz & W. Hill, *The Art of Electronics*, Cambridge University Press, Cambridge, 1983.
10.   J. Millman & A. Grabel, *Microelectronics*, McGraw-Hill International Book Company, 1987.
11.   J. Millman & C. C. Halkias, *Integrated Electronics: Analog and Digital Circuits and Systems*, McGraw-Hill Kogakusha Ltd., 1972.
12.   A.S. Sedra & K. C. Smith, *Microelectronic Circuits*, Holt, Rinehart and Winston, 1982.

Chapter 2

1.   R. King, *Integrated Electronic Circuits and Systems*, Van Nostrand Reinhold (UK) Co. Ltd., 1983.
2.   G. J. Ritchie, *Transistor Circuit Techniques*, Van Nostrand Reinhold (UK) Ltd., 1983.
3.   D. L. Schilling & C. Belove, *Electronic Circuits: Discrete and Integrated*, McGraw-Hill Kogakusha, Ltd., Tokyo, 1979.

Chapter 3

1. G. B. Clayton, *Operational Amplifiers*, Newnes-Butterworths, London, 1979.
2. L. M. Faulkenberry, *An Introduction to Operational Amplifiers*, John Wiley & Sons, NY, 1977.
3. M. S. Ghausi & K. R. Laker, *Modern Filter Design: Active RC and Switched Capacitor*, Prentice-Hall Inc., NY, 1981.
4. D. H. Horrocks, *Feedback Circuits and Operational Amplifiers*, Van Nostrand Reinhold (UK) Co. Ltd., 1983.
5. K. A. Key, *Analogue Computing for Beginners*, Chapman and Hall, London, 1965.

Chapter 4

1. T. C. Edwards, *Introduction to Microwave Electronics*, Edward Arnold Ltd., London, 1984.

Chapter 5

1. H. M. Berlin, *Design of Phase Locked Loop Circuits with Experiments*, Blacksburg Continuing Education Series, 1979.
2. R. E. Best, *Phase Locked Loops: Theory, Design and Applications*, McGraw-Hill Book Company, 1984.
3. F. M. Gardner, *Phaselock Techniques*, John Wiley & Sons Inc., NY, 1979.
4. D. F. Geiger, *Phaselock Loops for DC Motor Speed Control*, John Wiley & Sons, New York, 1981.

Chapter 6

1. J. A. Betts, *Signal Processing, Modulation and Noise*, The English Universities Press Limited, London, 1970.
2. J. J. O'Reilly, *Telecommunication Principles*, Van Nostrand Reinhold (UK) Ltd., 1984.

Chapter 7

1. D. B. Bruck, *Data Conversion Handbook*, Hybrid Systems Coroporation, 1974.
2. G. B. Clayton, *Data Converters*, The Macmillan Press Ltd., London, 1982.
3. P. H. Garrett, *Analog I/O Design, Acquisition: Conversion: Recovery*, Reston Publishing Company Inc., Virginia, 1981.
4. D. F. Hoeschele, *Analog-to-Digital/Digital-to-Analog Conversion Techniques*, John Wiley & Sons, Inc., NY, 1968.
5. Intersil, *Hot Ideas in CMOS*, 1983.
6. H. Schmid, *Electronic Analog/Digital Conversions*, Van Nostrand, New York, 1970.
7. A. H. VanDoren, *Data Acquisition Systems*, Reston Publishing Company, Inc., Virginia, 1982.
8. A. B. Wilkinson and D. H. Horrocks, *Computer Peripherals*, Hodder and Stoughton, London, 1980.

Chapter 8

1. L. O. Chua & P. M. Lin, *Computer-Aided Analysis of Electronic Circuits*, Prentice-Hall Inc., 1975.
2. J. A. Connely, *Analog Integrated Circuit Systems*, J. Wiley & Sons, 1975.
3. J. K. Fidler & C. Nightingale, *Computer Aided Circuit Design*, Thomas Nelson Ltd., 1978.
4. P. E. Gray, D. DeWitt, A. R. Boothroyd & J. F. Gibbons, *Physical Electronics and Circuit Models of Transistors*, J. Wiley & Sons, 1964.
5. G. J. Herskowitz & R. B. Schilling, *Semiconductor Device Modeling for Computer-Aided Design*, McGraw-Hill Book Company, NY, 1972.
6. M. F. Hordeski, *CAD/CAM Techniques*, Preston Publishing Company, USA, 1986.
7. R. L. Ramey, *Matrices and Computers in Electronic Circuit Analysis*, McGraw-Hill Book Company, NY, 1971.
8. A. V. Thompson, *MICRO-CAP II Microcomputer Circuit Analysis Program*, Spectrum Software, 1985.
9. J. Vlach & K. Singhal, *Computer Methods for Circuit Analysis and Design*, Van Nostrand Reinhold Company Inc., NY, 1983.

Papers

1. P. V. Ananda Mohan, V. Ramachandran & M. N. S. Swamy, Highpass and Bandpass 2nd-order Switched-Capacitor Filters, *IEE Proceedings*, **130**, Pt. G, No. 1, February 1983, pp. 1–6.
2. N. D. Arora, K. Nandkumar, N. H. Godhwani & K. C. Chhabra, Computer Analysis of Integrated Circuits, *Journal of Institution of Electronics and Telecommunication Engineers*, **24**, No. 12, 1978, pp. 517–525.
3. N. D. Arora, Computer-Aided Bipolar Transistor Model – Part 1: Development of Model, *Journal of Institution of Electronics and Telecommunication Engineers*, **25**, No. 8, 1979, pp. 345–348.
4. A. Badar & A. Basak, A Data Acquisition System for Magnetic Measurements, *Physica Scripta*, **39**, 1989, pp. 492–5.
5. A. Basak, A Constant Amplitude Oscillator, *Wireless World*, November 1969, p. 530.
6. A. Basak, A Low Frequency Waveform Sampling Unit, *International Journal of Electronic Engineering Education*, **12**, No. 3, 1975, pp. 322–7.
7. A. Basak & T. H. Al-Doori, Microprocessor Controlled Brushless DC Linear Motor, *IEE Conference Publication*, No. 234, May 1984, pp. 402–4.
8. G. R. Boyle, B. M. Cohn & D. O. Pederson, Macromodeling of Integrated Circuit Operational Amplifiers, *IEEE Journal of Solid-State Circuits*, **SC-9**, No. 6, December, 1974, pp. 353–364.
9. R.W. Brodersen, MOS Switched-Capacitor Filters, *IEEE Proceedings*, **67**, No. 1, January 1979, pp. 61–75.
10. Burr-Brown Research Corporation, USA, *Burr-Brown Instrumentation Amplifiers*, 1969.

11. J. G. Fossum, A Bipolar Device Modeling Technique Applicable to Computer-Aided Circuit Analysis and Design, *IEEE Transactions on Electron Devices*, **ED-20**, No. 6, June 1973, pp. 582–593.

12. S. Fotopoulos & T. Deliyannis, Multiple-loop-feedback Switched Capacitor Filters using Approximate Biquads, *IEE Proceedings*, **133**, Pt. G, No. 2, April 1986, pp. 84–8.

13. I. Getreu, Modeling the Bipolar Transistor: Part 1, *Electronics*, 19th September 1974, pp. 114–120.

14. I. Getreu, Modeling the Bipolar Transistor: Part 2, *Electronics*, 31st October 1974, pp. 71–75.

15. I. Getreu, Modeling the Bipolar Transistor: Part 3, *Electronics*, 14th November 1974, pp. 137–143.

16. G. Merckel, J. Borel & N. Z. Cupcea, An Accurate Large-Signal MOS Transistor Model for Use in Computer-Aided Design, *IEEE Transactions on Electron Devices*, **ED-19**, No. 5, May 1972, pp. 681–690.

17. J. E. Solomon, The Monolithic Op. Amp.: A Tutorial Study, *IEEE Journal of Solid-State Circuits*, **SC-9**, No. 6, December 1974, pp. 314–32.

18. G. Teague, Analysing Noise Errors in Op. Amps., *Electronic Product Design*, April 1982, pp. 53–56.

19. L. Yue & S. Y. Chao, SCNSOP: A Switched-Capacitor Circuit Simulation and Optimisation Program, *IEE Proceedings*, **133**, Pt. G, No. 2, April 1986, pp. 107–112.

Index

Printed in the United States
By Bookmasters